Space and Time, Matter and Mind

W. Schommers
KARLSRUHE, GERMANY

Space and Time, Matter and Mind

The Relationship between Reality and Space-Time

World Scientific
Singapore • New Jersey • London • Hong Kong

Published by

World Scientific Publishing Co. Pte. Ltd.

P O Box 128, Farrer Road, Singapore 9128

USA office: Suite 1B, 1060 Main Street, River Edge, NJ 07661

UK office: 73 Lynton Mead, Totteridge, London N20 8DH

SPACE AND TIME, MATTER AND MIND
The Relation between Reality and Space-Time

ISBN: 981-02-1851-6

Printed in Singapore by Uto-Print

Prologue

When we cogitate in physics about the nature of reality, we usually proceed in the following way: We develop within our mind a theoretical picture of reality, and then we prove it experimentally. If the result of such an experimental test is positive, it is usual to assume that the nature of reality is identical with the theoretical picture of reality. But we have to be careful.

As an example, let us consider the mechanistic picture of the world. Within this picture of reality, the world is a giant machine consisting of single objects which are connected to each other by single events: Material objects as the sun, moon and earth are embedded in space and they move under mutual influence according to deterministic laws. It belongs to the entity of these laws that the future and the past of mechanical systems are completely determined if their present state is known, i.e., *time* volatilizes, time is completely immured by the present instant; the future of a mechanical object can be predicted with absolute certainty and also his behaviour in the past can be retraced if its present state is known in detail.

This reality, described by classical mechanics, is primarily a product of the mind; it is a vision, and because of its large experimental success, most people actually believe that it is identical to the real world. But we experience concretely and directly only the vision, whereas the nature of reality outside remains merely vague, even when we are convinced of the opposite.

This assessment can also be read in a text by *Arthur Eddington* (1882-1944) (Swarthmore Lecture (1929), *Science and the Unseen World*):

"In comparing the certainty of things spritual and things temporal, let us not forget this - Mind is the first and most direct thing in our experience; all else is remote inference ..."

As is well known, classical mechanics is based on equations of motion (Newton's equations of motion), which have the form of differential equations, i.e., in connection with Eddington's and our remarks made above, the most direct "thing", which can teach us something about the nature of reality, is a system of differential equations which is located in our head or written on a sheet of paper; the complete evolution as well as the complete past of a physical system is given in the form of differential equations if the state at some time is sufficiently precisely known.

How reality outside is actually composed is at first only an assumption. The belief that reality outside has actually that composition which the theory preset, is entirely due to the success of the theory. For example, the discovery of the planet Neptune had produced an immense trust: Due to the irregularities in the motion of the planet Uranus, it was concluded that another planet should exist. It could be predicted precisely from the calculations "when" and in "what direction of space" one should look in order to observe its appearance. Le Verrier in Paris and Adams in London made these calculations, and then Le Verrier wrote in 1846 a letter to a colleague in Berlin with the instruction to look in a certain direction at a certain time, and the colleague in Berlin actually observed the new planet Neptune. *"An astronomer had discovered a new planet without raising his head."* [1]

However, what we observe with our eyes or with a telescope is only a picture, a hypothesis of reality. This is because the signals of reality outside that we ingest via our sense organs are processed (unconsciously) by the brain, giving the impression of what we have directly in front of us, and we are firmly convinced that this impression

(picture) is identical with the situation of reality outside. For example, the well-known psychologist *C.G. Jung* (1875 - 1961) wrote in [2]:
"When one thinks about what consciousness really is, one is deeply impressed of the wonderful fact that an event that takes place in the cosmos outside, produces an inner picture, that the event also takes place inside ... "
(*"Wenn man darüber nachdenkt, was das Bewußtsein wirklich ist, ist man zutiefst von der wunderbaren Tatsache beeindruckt, daß ein außerhalb im Kosmos stattfindendes Ereignis gleichzeitig ein inneres Bild hervorruft, daß das Ereignis sozusagen ebenso im Inneren stattfindet ... "*)

But is the inner event identical with that in reality outside? As already mentioned, what we see in front of us is an inner event even when we have the impression that everything is arranged outside. But there are clear indications of the fact that the *picture inside* is not identical with the *event outside*. With the assumption that reality is composed in such a manner as we see it in front of us we probably deceive ourselves as those who believe that the virtual picture behind a mirror is a material body. In this monograph, we will discuss this point in detail.

The situation can, therefore, be understood as follows: The signals from reality are processed by the perception system of the brain into a picture using certain instructions and rules, and it is important to note that these rules and instructions are primarily not constructed by the features *true* and *untrue* but by a *principle of usefulness*, i.e., to distinguish between good and bad conditions of life. In other words, in connection with the development of the perception system by the evolutionary processes it was primarily not necessary that an individual knows how reality outside is actually composed but it was essential for the individual to *survive* in his environment.

We can also say that *signals of reality* are processed by the perception system into a *picture of reality* on the basis of a theory, where the theory contains specific, in particular, species-preserving instructions; it is a theory which comes into play over the biological apparatus without the deliberate agency of the observer.

Therefore, if we compare results, which are governed by Newton's mechanics, with situations which are directly in front of us or which we observe through a telescope, we actually compare a theory with a theory, but not a theory with the actual reality that exists outside, indepedent of the observer. This is in contradiction to the above statement by C.G. Jung, and most people are firmly convinced that the conception reflected in Jung's statement is true.

In conclusion, the question formulated above, whether the composition of reality outside is identical with the features of a sufficiently experimentally-tested world view, must be answered in the negative.

What are the differences between a certain world view and reality outside? How exact is the correspondence between them? These and other questions have to be answered.

The composition of reality, so as we have it, for example, in front of us in everyday life, is in any case dependent on the observer, even then when we assume that it exists outside in exactly the same composition. This is because the signals coming from the reality outside are always processed into a picture by the observer, and the observer belongs in any case to a certain biological species. How could one show or prove that the features of reality outside are identical with those of the picture? We need a diversity of mechanical systems in order to test extensively Newton's mechanics, so we need a diversity of biological systems in order to prove whether a general conception of reality exists, i.e., whether an observer-independent reality is perceptible. To this set of problems, very interesting experiments have been performed, which lead to the conclusion that reality outside - i.e., an observer-independent reality - is not perceptible.

This fact should be considered in the formulation of physical laws, and this has to be done from the start. Before we formulate a physical law, we should have a conception of how the terms "reality" and "observation" are arranged. *Robert Havemann* (1910-1982) expressed an important principle in physics as follows [3]: *"What we think about must deal with that what really exists in reality."* (*"Was wir denken, muß zum Inhalt haben, was in der Wirklichkeit existiert."*) How can we fulfil this principle when the actual reality is not perceptible?

Havemann´s formulation involves that the conceptions, objects, etc., in natural science must have a counterpart in reality outside, i.e., only such elements and quantities are accepted which can be proved by measurements. But does there really exist a theory which fulfils this demand? Hardly, because the physics is teeming with *metaphysical elements*, i.e., there are quantities and conceptions in physics which are in principle not accessible to measurements. *Gottfried Falk* (1922-1991) stated in [4] the following:
" ... highly assessed statements like the two basic laws of thermodynamics or the assertion that matter is once and for all built up by "last building blocks", the elementary particles, are not scientific, but metaphysical."
(*" ... hoch bewertete Aussagen wie die beiden Hauptsätze der Thermodynamik, oder die Behauptung, daß die Materie aus ein für alle mal, also absolut gegebenen "letzten Bausteinen", eben den Elementarteilchen, zusammengesetzt ist, nicht naturwissenschaftlich, sondern metaphysisch sind."*)

After Falk´s criteria quoted in [4], it is questionable whether a system of physics can be set up at all in which all the elements have measureable counterparts in reality outside. If, however, the picture of reality is principally not identical with the composition of reality outside (a feature which has been discussed above) such a demand loses sense since the actual reality is in principle not accessible to the observer, and this is because an observer-independent reality obviously cannot be perceived; in such a situation we do not compare a *theory with an experiment*, but a *theory with a theory*; in such a case the criteria of assessment will be steered to features as *inner consistence* and *logical structure*.

We know from *quantum theory* that a phenomenon is always an *observed* phenomenon [5] (Copenhagen interpretation of quantum theory); thus it is senseless to talk from a phenomenon at all without observation. In other words, within the frame of the Copenhagen interpretation of quantum theory, the observer plays a fundamental role. But this fact is completely different from that which we have said above about the observer-independent reality: This reality exists even when we cannot perceive it; within quantum theory, reality can, however, only be produced in connection with an observation. Furthermore, within the frame of the Copenhagen interpretation of quantum theory, the observer is involved very generally but not his specific composition, and this is quite different from that which we have stated above in connection with the observer-independent reality.

As mentioned above, the fact that an observer-independent reality cannot be perceived has to be considered from the beginning in the formulation of physical laws. Before we formulate physical laws we should have a conception of how the terms "reality" and "observation" are arranged. This consideration must lead to new conceptions and features for classical mechanics as well as for quantum theory. Important terms and principles have to be newly thought-out, rejected, modified, etc., and this should be the case up to fundamental terms like space and time. A more detailed discussion of this point is given in [5].

In order to be able to include the above-discussed tendencies, it is in our opinion essential to know the

relationship between reality and space-time.

The purpose of this monograph is to investigate this point in detail; in this connection the role of

matter and mind

will be examined extensively.

We will start our discussion with a critical consideration of space and time, where *Mach's principle* will be the focus of attention. After that, we will bring indications and proofs of the idea that the picture of reality is not identical with reality outside. In particular, the difference between the *picture* and *reality outside* will be specified by the inclusion of space and time. The new results will be substantiated through facts from the philosophy of science.

I would like to thank E. L. Haase and M. Rieth of the Kernforschungszentrum Karlsruhe and C. Politis of the University of Patras for many helpful remarks on the material in this book. I am indebted to many people for carefully reading and checking the manuscript. I am particularly grateful to Miss Gillian Chee (Editor) of World Scientific for her cooperation in the production of this book.

CONTENTS

Chapter 2 - The Point of View of Philosophy of Science

Chapter 1

Levels of Reality

1.1 Space-Time Conceptions

The methodological action in today´s physics is strongly influenced by the conceptions introduced by *Galilei* (1564-1642), *Kepler* (1571-1630) and *Newton* (1643-1727). By *mathematical thinking* and by *systematically performed experiments*, they established classical mechanics and astronomy, respectively. In particular, with Newton´s "Principia" a new interpretation-scheme for natural phenomena has been introduced which is valid up to the present day: Newton investigated the laws of nature by *abstract analysis*, i.e., by the separation of isolated problems out of the overall context of natural phenomena.

In this introductory chapter, we would like to discuss Newton´s conception of space and time and show how this space-time conception has been changed by the *theory of relativity*.

1.1.1 Newton´s Mechanics

Where does the inertia come from?

In Newton´s physics, space and time are *absolute quantities*, they are independent from each other, and they may even exist when space is not filled with matter. Only with these space-time features Newton was able to construct a reasonable theory of motion.

However, the concept of an absolute space (and that of an absolute time) has led to enormous intellectual difficulties which - as we will see below - could not be eliminated at all by the theory of relativity. In the following sections,we would like to discuss in more detail why the concept of absolute space causes serious problems.

Concerning the term "absolute", note the following:

1. Absolute space was invented by Newton for the explanation of *inertia*. However, we do not know of any other phenomenon for which absolute space would be responsible. So, the hypothesis of absolute space can only be proved by the phenomenon (inertia) for which it has been introduced. This is unsatisfactory and artifical.

2. The term "absolute" not only means that space is *physically real* but also *"independent in its physical properties, having a physical effect, but not itself influenced by physical conditions"*[6]. This must also be considered as unsatisfactory.

Both points indicate that absolute space is actually an *unphysical* quantity. Although Newton´s mechanics was very successful (and it is still used in many calculations) a lot of physicists could not accept the concept of an absolute space. This is demonstrated by the fact that scientists tried to solve this problem again and again up to the present day.

Concerning absolute space, *Max Born* (1882 - 1970) expressed his view very clearly. He wrote [7]:

"Indeed, the concept of absolute space is almost spiritualistic in character. If we ask, "What is the cause of centrifugal forces?", the answer is: "Absolute space." If, however, we ask what absolute space is and in what other way it expresses itself, no one can furnish an answer other than that absolute space is the cause of centrifugal forces but has no further properties. This consideration shows that space as the cause of physical occurrences must be eliminated from the world picture."

("Der absolute Raum aber hat nahezu spiritistischen Charakter. Fragt man: "was ist die Ursache für die Fliehkräfte?", so lautet die Antwort: "der absolute Raum". Fragt man aber: "was ist der absolute Raum und worin äußert er sich sonst?", so weiß niemand eine andere Antwort als die: "der absolute

Raum ist die Ursache der Fliehkräfte, sonst hat er keine Eigenschaften". Diese Gegenüberstellung zeigt zu Genüge, daß der leere Raum als Ursache physikalischer Vorgänge aus dem Weltbild beseitigt werden muß.")

Furthermore, we find in connection with this subject [7]:

"Sound epistemological criticism refuses to accept such made-to-order hypothesis. They are too facile and are at odds with the aim of scientific research, which is to determine criteria for distinguishing its results from dreams of fancy. If the sheet of paper on which I have just written suddenly flies up from the table, I should be free to make the hypothesis that a ghost, say the spectre of Newton, had spirited it away. But common sense leads me instead to think of a draft coming from the open window because someone is entering by the door. Even if I do not feel the draft myself, this hypothesis is reasonable because it brings the phenomenon which is to be explained into a relationship with other observable events."

("Eine gesunde Erkenntniskritik lehnt solche ad hoc gemachten Hypothesen ab; sie sind zu billig und zerbrechen alle Schranken, die gewissenhafte Forschung zwischen ihren Ergebnissen und den Hirngespinsten der Phantasie aufzurichten versucht. Wenn der Bogen Papier, den ich eben beschrieben habe, plötzlich vom Tisch auffliegt, so stände mir die Hypothese frei, daß der Geist des längst verstorbenen Newton ihn entführt habe; aber als vernünftiger Menschen mache ich diese Hypothese nicht, sondern denke an die Zugluft, die entstand, weil das Fenster offensteht und meine Frau gerade zur Tür hereintritt. Auch wenn ich die Zugluft nicht selbst gespürt habe, ist diese Hypothese vernünftig, weil sie den zu erklärenden Vorgang mit einem anderen beobachtbaren in Verbindung bringt.")

In conclusion, the concept of absolute space (and of course that of absolute time) is problematic and such a concept should not have a place in physical theories. Nevertheless, such theories have been tolerated because they were very successful in the description of experiments. This is not only true for Newton's mechanics but also - as we will see below - for the special and general theory of relativity.

What are space and time made of?

How can we understand Newton´s absolute space which can even exist without the existence of material bodies? Sure, it is the seat of inertia. But what is it? What is it made of? It is made of nothing! Such a construct cannot be taken as the foundation of a scientific theory of motion. Newton closed this logical gap by providing the empty space with divine attributes. So, Newton combined - against his own principles - *physical* with *metaphysical* arguments.

Since Galilei the tendency is predominantly not to use *dogma* (i.e., a system of beliefs, put forward by some authority to be accepted as true without question) in the scientific treatment of phenomena in nature. One reason why Newton´s absolute space and absolute time are *dogmatic* in character is that they should reflect the divine omnipresence.

Therefore, Newton´s conception of space and time has been criticized immediately by influential philosophers as *Berkeley* and *Leibnitz*. But the big success of Newton´s mechanics overshadowed these doubts.

1.1.2 Mach´s Principle

It was not until the end of the nineteenth century that the concepts space and time were analysed once again, in particular by *Ernst Mach* (1838-1916). Mach demanded in a radical manner the elimination of all metaphysical elements - like *absolute* space and *absolute* time - from the building of physics.

The elimination of space as an active cause

This is why Mach eliminated space as an active cause in the system of mechanics (Mach's principle). According to him, a particle does not move in unaccelerated motion relative to space (as in the case of Newton's conception), but relative to the centre of all the other masses in the universe [6]; in this way, the series of causes of mechanical phenomena was closed, in contrast to Newton's mechanics.

In fact, absolute space and, of course, absolute time must be considered as metaphysical elements because they are, in principle, *not accessible to empirical tests:*

There is no possibility of determining space coordinates x_1, x_2 , x_3 and time t. We can only say something about distances in connection with masses, and times in connection with physical processes.

Even if one does not principally reject metaphysical elements, one has to recognize that in the case of space-time, it would be made too easy for oneself, because the above-mentioned fact that space and time can only be observed in connection with matter could possibly reflect a fundamental principle. Even if one is generally not a follower of Mach's philosophy, one has to take *Mach's principle* (formulated above) literally, because we have to accept that the space-coordinates and the time are not observable as independent quantities, i.e., their absolute character, which these quantities have within the framework of Newton's mechanics, is in contradiction to experience!

Mach's writings had a strong influence on *Einstein*; with the publication of the *special and general theory of relativity*, new space-time conceptions were created.

1.1.3 Conception of Relativity

Relativity of inertia

Within the special theory of relativity (STR), space and time are no longer independent of each other; they are tied together into a space-time. Is Mach's principle fulfiled within the STR? Definitely not: Newton's three-dimensional space is merely extended within the STR to a four-dimensional space-time, without overcoming the absoluteness. In other words, instead of Newton's absolute space, within the STR we have an absolute space-time (Minkowski's space); also this space is - as in Newton's three-dimensional space - the seat of the absolute forces of inertia. In order to *relativize* the forces of inertia Einstein was led to formulate the general theory of relativity (GTR). In this way Einstein wanted to eliminate the absolute space-time, which can only be different from nothing in connection with metaphysical elements (for example, divine attributes as in the case of Newton's mechanics). In accordance to Mach, Einstein argued that the inertia of an object should be completely due to the other matter in the universe; this relativity of inertia was the foundation of his entire considerations.

With the formulation of the GTR, Einstein took up a completely new direction; this theory represents a magnificent building, and its results are confirmed by many experiments.

However, the GTR failed to fulfil its initial goal, namely to eliminate the *absolute space-time* of the STR. We would like to discuss this point by means of some solutions which follow from Einstein's field equations.

de Sitter's universe

In 1917, *de Sitter* (1872-1934) gave a solution to Einstein's equations which corresponds to an empty universe, i.e., within the framework of this solution, space-time could exist without matter, and this is in obvious contradiction to Mach's principle. This fact was annotated by Einstein's collaborator *Banesh Hoffmann* as follows [8]:

"Barely had Einstein taken his pioneering step when in 1917 in neutral Holland de Sitter discovered a different solution to Einstein's cosmological equations. This was embarrassing. It showed that Einstein's equations did not lead to a unique model of the universe after all. Moreover, unlike Einstein's universe, de Sitter's was empty. It thus ran counter to Einstein's belief, an outgrowth of the ideas of Mach, that matter and space-time are so closely linked that neither should be able to exist without the other."

Here, a space-time without material objects can also only be nothing - as within Newton's theory.

Gödel's solution

Another solution of Einstein's field equations was presented by *Gödel* (1906-1978) in 1949 [9]. Gödel's solution and his significance was very clearly discussed by Heckmann [10]:

"This solution by Gödel describes a model of the world which is uniformly filled with matter. All points in it are equivalent, which is therefore homogeneous as in the cases mentioned up to now; it is infinitely large and rotates absolutely, but is not able to expand. At the beginning, in 1916, and still for a long time after that, Einstein himself believed that his theory would contain the relativity of all motions. Gödel's solution was the first solid evidence that Einstein's belief was an error."

("Diese Gödel-Lösung beschreibt ein mit Materie gleichförmig erfülltes Weltmodell, dessen Punkte alle gleichwertig sind, das also homogen ist, wie die bisher von uns erwähnten Fälle, das dabei unendlich ausgedehnt ist, das aber absolut rotiert, jedoch nicht expandiert. Das Überraschende ist die absolute Rotation. Einstein selbst hatte anfangs, also um 1916, und auch noch lange Zeit nachher geglaubt, seine Theorie enthalte die Relativität aller Be-

wegungen. Die Gödelsche Lösung war der erste harte Beweis dafür, daß die-
ser Glaube Einsteins ein Irrtum war".)

In conclusion, within the framework of the GTR we can develop conceptions which still contain the concept of absolute space. The absoluteness of space in the sense of Newton is therefore still completely
contained within Einstein´s theory.

In the primary formulation of the field equations of the GTR, Mach´s
principle has obviously not been considered. But there is still the possibility to include this important principle into the GTR by suitable
procedures. In the following section, we would like to investigate this
point.

1.1.4 Can Mach´s Principle Completely be Realized?

As is well known, within the framework of the STR, the square of the
invariant distance ds of the world lines in the four-dimensional space
is given by

$$ds^2 = dx_1^2 + dx_2^2 + dx_3^2 + dx_4^2 \ ,$$

(1)

where

$$dx_4^2 = -c^2 dt^2 \ .$$

(2)

x_1, x_2 and x_3 are again the space-coordinates and t is the time. Within the GTR, curved coordinates are used, and in this case the distance
ds is generally expressed by

$$ds^2 = \sum_{i,k=1}^{4} g_{ik} dx_i dx_k \ ,$$

(3)

where

$$g_{ik} = g_{ki} \tag{4}$$

can be put. The geometry which is expressed by (3) is in general non-euclidean; the coordinate x_4 can in general no longer be identified with time t, and coordinates x_1, x_2 and x_3 in general no longer with the space-coordinates.

The ten metric coefficients g_{ik}, which are after Einstein, completely determined by the distribution of the material bodies, have - as is well known - a double-entry function:
1. They fix the *metric*, i.e., the units of distance and time.
2. They express the *gravitational field* of the usual mechanics.

Geometry and *gravitational field* are within the framework of the GTR two aspects of the same thing; both are decribed by the ten coefficients g_{ik}.
In the case of

$$g_{11} = g_{22} = g_{33} = g_{44} = 1 ,$$
$$g_{ik} = 0 ; i \neq k , \tag{5}$$

Eq. (3) is identical with (1), i.e., we obtain the result of the STR (Minkowki´s space). Deviations from (5) reflect a state which is called in Newton´s mechanics gravitation; the motions of inertia are curved and are not uniform, and for such a state within Newton´s mechanics, attractive forces are responsible.

Since we always have to identify the geometry with a field within the GTR, case (5) also represents a field; it is a gravitational field with a curvature of zero. The motion within such a field is a motion of inertia with constant velocity. After the principle of Mach, this field must also have its origin in other masses; it may not be a feature of space.

Now let us come to the essential point: Are there any solutions within the GTR which fulfil *Mach's principle* in the above sense? Are, in other words, the motions of inertia and centrifugal forces, which are experienced by bodies, *completely* due to other masses in the universe?

On the basis of the above-discussed cases, for example, *Gödel's solution*, we have seen that Einstein's field equations still contain the concept of *absolute space*; in other words, Mach's principle cannot be a consequence of Einstein's field equations. But there is perhaps the possibility to fulfil Mach's principle by an extension of Einstein's equations by means of the introduction of additional elements. This is obviously not the case or - if at all - only in connection with serious modifications. Concerning this point let us quote the following remarks by Born [7]:

"Concerning Mach's principle, which was Einstein's starting point, we have mentioned in section 9 on page 296 that THIRRING could prove that centrifugal forces and forces of inertia are due to the rotation of masses which are far away. However, with that it could only be shown that masses, which are far away, have - in accordance with Mach's principle - an influence on a test-body at all, but with that not all forces of inertia were explained. This is because in THIRRING'S work these masses are in a world which has at large distances from all masses Minkowki's metric [see Eq.(99); identical with (5) in the present text]. But the complete principle of Mach requires that in a region of space, which is free of matter, Minkowki's metric should also be produced by masses. In conclusion, within a complete empty universe, Minkowki's g_{ik} (99) [identical with (5) in the present text] should not be valid but all coefficients g_{ik} should be given by $g_{ik}=0$. Then all the motions of a test-body would occur without forces of inertia, as required by Mach's principle ..."

("Was nun das Mach'sche Prinzip anlangt, von dem Einstein ausgegangen

war, so haben wir in Abschnitt 9 Seite 296 erwähnt, daß THIRRING die Er-
zeugung von Zentifugal- und anderen Trägheitskräften durch die Rotation
ferner Massen nachgewiesen hat. Damit ist aber nur gezeigt worden, daß
ferne Massen überhaupt Wirkungen auf einen Probekörper entsprechend
dem Mach´schen Prinzip ausüben, nicht aber sind alle Trägheitskräfte da-
mit vollständig erklärt. Denn die fernen Massen befinden sich in
THIRRING`s Arbeit in einer Welt, die in großen Entfernungen von allen
Massen die Minkowkische Metrik hat (siehe Gl.(99) (identisch mit Gl.(5)
in dem hier vorliegenden Text). *Nach dem vollständigen Mach´schen*
Prinzip sollte aber in einem Raumgebiet, das selbst materiefrei ist, auch die
Minkowskische Metrik selbst, ... , durch die Verteilung ferner Massen erst
erzeugt werden. Es sollten also im gänzlich leeren Universum nicht etwa die
Minkowskischen g_{ik} (99) (identisch mit (5) im vorliegenden Text) *gelten*
sondern es sollten alle $g_{ik}=0$ sein. Dann würden in der leeren Welt beliebige
Bewegungen eines Probekörpers ohne Trägheitskräfte verlaufen, wie es das
Mach´sche Prinzip verlangt...")

The existence of Minkowki´s metric means that within the GTR the
test-body experiences forces of inertia, completely in analogy to
Newton´s mechanics. This is because within the framework of the
GTR, geometry and gravitation are the same things, as we have already mentioned above, and the geometry with Minkowki´s metric
corresponds to a gravitational field with a curvature of zero. Since in
Thirring´s case this metric cannot be derived from other masses, we
have to conclude that these forces of inertia must have their origin in
space which is therefore *absolute* in character, completely in analogy
of Newton´s mechanics. As Born correctly remarked (see the text
above), this feature is reflected by the fact that condition (condition for
Mach´s principle)

$$g_{ik} = 0 \; ; \; i,k = 1,...,4$$

(6)

is not fulfiled but (5).

In conclusion, within Thirring's analysis, Mach's principle is only incompletely realized. The question whether Mach's principle can *completely* be fulfiled within Einstein's theory seems to be an open question. Although there are some interesting discussions in the literature, the problem can be considered as not solved, even not approximately. In the opinion of many authors, the programme to incorporate Mach's principle into the GTR has not been carried out. For example, in [11] the following is quoted:

"Although Einstein was led in the construction of his theory by Mach's idea of the ontological dominance of matter over space-time, he was even convinced temporarily that his theory "had taken space and time the slightest trace of an objective reality" (100, S. 831). But the opposed tendency turned out to be the case: A tendency against Mach's principle became effective by a kind of an inherent dynamism of the theory in the course of time of the development of the conception. Space-time became more and more an ontologigally respectable entity, and it took over more and more the properties of material objects."

("Obwohl Einstein bei der Konstruktion seiner Theorie geleitet war von der Machschen Idee der ontologischen Dominanz der Materie über die Raumzeit, ja sogar zeitweise glaubte, daß seine Theorie "Raum und Zeit die letzte Spur einer objektiven Realität genommen hatte" (100, S. 831), zeigte sich durch eine Art Eigendynamik der Theorie im Laufe der Ausarbeitung der Konzeption die gegenläufige Tendenz, anti-machsche Tendenz. Raumzeit wurde immer mehr zu einer ontologisch respektablen Entität, und sie übernahm laufend Eigenschaften von materialen Objekten.")

The relationship between the GTR and Mach's principle has also been discussed instructively by *Dehnen* [12]:

"In those days Einstein had in mind that the structure of space and time is given completely by the particular distribution of matter in the world in accordance with his field equations of gravitation. As a result of this, Mach's idea would be fulfiled simultaneously, after which the inertia of material

bodies is determined by other masses in the world ... However, it should be
emphasized that Einstein's vision that Mach's principle could be realized
within the framework of the general theory of relativity failed, even by an additional modification of the original field equations ... The problem in connection with the absolute space-time within the framework of the special theory of relativity - a relic from Newton's mechanics - is, in the general theory
of relativity, still not solved ... "

("Dabei schwebte Einstein seinerzeit selbst vor, daß die Struktur von Raum
und Zeit durch die jeweilige Materieverteilung in der Welt nach Maßgabe
seiner Feldgleichungen der Gravitation bedingt und restlos bestimmt wird.
Hierdurch würde zugleich der Machschen Idee Rechnung getragen, wonach
das Trägheitsverhalten materieller Körper durch die Massen in der Welt
festgelegt wird ... Es ist jedoch hervorzuheben, daß sich diese Einsteinsche
Vision des "Mach'schen Prinzips" im Rahmen der Allgemeinen Relativitätstheorie nicht hat realisieren lassen, selbst nicht nach zusätzlicher Modifi-
kation der ursprünglichen Feldgleichungen ... Die Problematik der speziell-
relativistischen absoluten Raum-Zeit - ein Relikt noch aus der Newtonschen
Physik - ist also auch in der Allgemeinen Relativitätstheorie gar nicht end-
gültig überwunden ...")

Dehnen also discusses different reasons why that is the case. One reason is obviously that Einstein's field equations are non-linear differential equations, "so that the metric field becomes - due to his self-interactions
- a certain independence from matter " ("so daß das metrische Feld infolge
Selbstwechselwirkungen eine gewisse Eigenständigkeit gegenüber der Materie erlangt") [12], and this obviously explains why the concept of absolute space is still a property of the GTR.

Serious modifications

In order to fulfil Mach's principle, serious modifications are obviously
necessary. This possibly leads to the situation that we have to leave

the actual frame of the GTR, for example, in the case where the gravitational constant G varies with time by the use of an additional constant of nature. This point is discussed in [13] as follows:

"The idea that the strength of gravity varies is the basis of a fully relativistic theory developed principally by Brans and Dicke. The theory involves space-time and geodesics, and generates uniform model universes with the Robertson-Walker separation formula (6.1.3). However, the dynamical equations connecting matter with the curvature of space-time are no longer the "simplest possible", as in Einstein´s theory, but involve an extra constant of nature, ω, whose effect is to make G vary with time. The introduction of ω was not an ad hoc complication of general relativity; it was hoped that the Brans-Dicke theory would satisfactorily incorporate Mach´s principle into relativity (the extent to which Einstein´s theory does this is still uncertain)."

1.1.5 Conclusions

Mach´s principle, which should be satisfied in any space-time theory, is not or is only incompletely fulfiled in Newton´mechanics and in the theories of relativity. Even the solutions of Einstein´s field equations still contain the concept of absolute space as a physical element.

Hypothetic theories of the Brans-Dicke type (see also [13]), which use fundamental new elements in order to be able to consider Mach´s principle, are already far away from the actual nucleus of the theory of relativity. Furthermore, the numerical value of the extra constant of nature w is so small that it is not possible to distinguish experimentally between Einstein´s theory and the theory of Brans-Dicke.

It is principally not very useful to propose theories in which Mach´s principle is only partly realized; in these cases, absolute space is not *completely* eliminated but only partly.

1.1.6 Summary

Let us summarize: Strictly speaking, Einstein's main goal to eliminate the absoluteness of space-time was not carried out.

What then are the changes in the concepts of space and time from Newton until today?

1. Space and time are no longer independent of each other; within the frame of the STR they are fused into a four-dimensional space-time continuum.
2. Within the GTR, geometry is in general no longer euclidean.

As mentioned, the *absolute character* of space-time could only be partly eliminated. Heckmann remarked [10]:
"The absoluteness of space, which Newton claimed, and Einstein thought to have eliminated, is still contained in Einstein's theory, insofar as in Einstein's theory the concept of an absolute rotation is completely legitimate. Within Einstein's theory, it is still possible to talk about the rotation of the entire matter of the world relative to absolute space."
("*Die Absolutheit des Raumes, die Newton behauptete, die Einstein aber glaubte aufgehoben zu haben, ist in der Einsteinschen Theorie noch völlig enthalten, insofern als der Begriff der absoluten Rotation auch in der Einsteinschen völlig legitim ist. Ja, es ist in Einsteins Theorie noch sinnvoll, von der Rotation des gesamten Materie-Inhalts der Welt relativ zum absoluten Raum zu sprechen*".)

For example, Gödel's solution of Einstein's field equations leads to a cosmos in which - if it would really exist in this form - the concept of an absolute space would actually be realized. However, this cannot be the case since absolute space must be considered as an unphysical concept.

Concerning absolute space, Dehnen remarked [12]:

"The problematic nature of Newton´s absolute space, which was emphasized by E. Mach and which has only been shifted from three to four dimensions in the transition to the special theory of relativity, has been partly eliminated by the general theory of relativity though not abolished."

("Die Problematik des Newtonschen absoluten Raumes, auf die E. Mach so nachdrücklich hingewiesen hat und die beim Übergang zur Speziellen Relativitätstheorie nur vom 3-Dimensionalen ins 4-Dimensionale verlagert wurde, wird eigentlich erst durch die Allgemeine Relativitätstheorie entschärft, wenn auch nicht aufgehoben.")

As already mentioned above, it is not very useful when Mach´s principle is only partly realized, because in such a case absolute space-time is not eliminated.

1.1.7 Other Conceptions

Without a doubt, the absolute character of space, which is completely developed within Newton´s theory and the STR, is partly eliminated within the GTR; within the GTR, distances and time-intervals are not independent from the contents of space, i.e., from the distribution of matter.

For example, the length of a yardstick which is situated near the mass M (see Fig. 1) is dependent on the distance relative to M, and this effect is described by (see, for example, [14])

$$l_1 = l_0\left(1 - \frac{\Delta\Phi}{c^2}\right) , \tag{7}$$

where $\Delta\Phi$ is the difference of the gravitation potential, and l_0 is the length of the yardstick for an infinite distance from M.

The test-body effect

The effect which we have discussed in connection with (7) has nothing to do with another feature, namely the fact, that distances, i.e., also l_0 and l_1 in (7), can only be observed in connection with material bodies (and time-intervals only in connection with physical processes) which have nothing to do with the mass M in (7). In other words, in the observation of the distances l_0 and l_1, which - due to the presence of M - are different from each other, we need material objects (test-bodies), where the mass of these material objects are primarily not of interest, i.e., they may be used for the determination of l_0 as well as for l_1. In the following, we would like to call this effect the *test-body effect*.

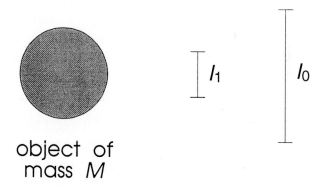

object of
mass M

Figure 1. Distances can only be observed in connection with material objects. Within the GTR, distances actually depend on the matter in space. For example, the length of a yardstick is dependent on the distance relative to M; according to the distance from M we measure l_0 and l_1, respectively. Of course, this does not also mean that the distances l_0 and l_1 can be measured with M; for this purpose we need additional material objects which we call *test-bodies* in the text.

Any space-time theory should take the test-body effect into account. Even if Mach´s principle were satisfied within the usual space-time theories, there would be no consideration of the test-body effect, i.e., the space-coordinates $x = x_1$, $y = x_2$, $z = x_3$ and time t remain metaphysical elements, because no test-bodies were used in their definition.

Is matter embedded in space-time?

Our views of the world are based on the conception that the material objects are *embedded* in space-time. Within the frame of such a conception, it is quite natural to assume the existence of an *empty* space-time. This often leads to the following expectation: Since we are able to observe the "elements of matter" (localized masses) independent from each other, we expect intuitively that we should also be able to observe the elements of space-time, the coordinates x, y, z and time t, independent from other coordinates $x´$, $y´$, $z´$ and other times $t´$. However, this conception is obviously wrong because it is in contradiction to empirical facts: We never observe the quantities x, y, z and t in isolation but only in connection with other coordinates $x´$, $y´$, $z´$ and times $t´$, i.e., we observe only distances and time-intervals.

What are the roots of this false conception? It might be due to the idea that we consider space as a certain kind of "substratum" which is filled with matter and fields. This is because the assumption that all real objects are embedded in space-time seems to implicate that space-time itself is a real something which we have called "substratum". Then, there are no difficulties in imagining taking objects out of the space and putting objects into the space, respectively. In particular, it seems to be almost natural that such a "substratum" can also exist in an *empty* state. In other words, such a space-time substratum can be considered as an independent quantity and its elements x, y, z and t

should be real physical quantities, analogous to the mass of an object.

However, as already mentioned above, this conception is obviously wrong, and this is due to the fact that we never observe the quantities x, y, z and t in isolation form but only in connection with other coordinates x', y', z' and times t', i.e., we observe only distances and time-intervals. Furthermore, we have outlined above that the distances and time-intervals can only be observed in connection with test-bodies and physical processes, respectively. This is strange and leads to the following question: May we really assume that matter is *embedded in* space-time? We possibly have to revise this view. This is because such a conception is somehow connected with the imagination that space-time itself represents a real something which we have called "substratum". But we have to consider space-time as a "nothing" because the elements x, y, z, t of such a "substratum" can in principle not be observed, i.e., they are as a real something not existent, and the concept of an *empty* space breakes down; the space (empty or full of matter) must therefore be considered as a "nothing" (see also the discussion in Subsec. 1.1.1). But how should a "nothing" be able to take up material objects and fields?

Many discussions in the literature indicate that the conception of an *empty* space should actually be considered as a prejudice. For example, Falk and Ruppel remarked [15]:

"But nevertheless we are accustomed to consider space as something special, namely as a substratum, in which all physical things, as objects and fields, are embedded. Space is the house which takes up physical objects, so to speak; it forms the stage where the processes take place. It seems clear to everybody what one means when he speaks about empty space, i.e., space which is free of matter and fields. But is that really the case? It is astonishing again and again how easily we adopt conceptions and how naturally and inevitabley we hold them. Certainly our conception of empty space belongs to this ca-

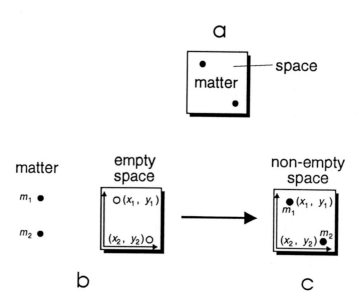

Figure 2. a. The assumption that matter is embedded in space-time seems to imply that space-time itself is a real something which we have called "substratum" in the text. Then we have no difficulties in imagining taking objects out of the space and putting objects into the space, respectively. In particular, it seems to be almost natural that such a "substratum" can also exist in an *empty* state. b. Then, we expect that the substratum in the empty state (the empty space) can be *marked*, i.e., we should be able to place coordinates. In the two-dimensional representation here, the two marker points have the coordinates (x_1, y_1) and (x_2, y_2). We obtain Fig. c, which is identical to Fig. a, when we embed the masses m_1 and m_2 into the empty space at the marker points. But this view is obviously wrong! The empty space cannot be marked, i.e., the situation which is given by Fig. b is not realistic; isolated coordinates are not observable but only distances in connection with material objects. The empty space is a "nothing" and cannot be identified with a "substratum", i.e., a real something. The conception that matter is *embedded in* space-time therefore becomes problematic.

tegory."

("*Aber dennoch sind wir gewohnt, den Raum als etwas Besonderes zu be-
trachten, nämlich als das Substrat, in das alle physikalischen Dinge, wie
Körper und Felder, eingebettet sind. Der Raum ist sozusagen das Haus, das
die physikalischen Objekte aufnimmt; er bildet die Bühne, auf der sich die
Vorgänge abspielen. Es scheint jedermann klar zu sein, was er meint, wenn
er vom leeren Raum, d.h. von Körpern und Feldern entblößten Raum
spricht. Aber ist das wirklich so? Es ist immer wieder erstaunlich, wie leicht
uns manche Vorstellungen eingehen und für wie selbstverständlich und
zwangsläufig wir sie halten. Unsere Vorstellung vom leeren Raum gehört si-
cher dazu.*")

Concerning time, the following remarks are instructive [16]:

"*As mentioned, the concept of time cannot actually be understood. We are
accustomed to think that time is something which can be found or which one
has. But it has actually no existence. The physicist says that time is something that can be measured in one or another way by a clock. But what does
the clock measure? Just the time!*

*Regarding this problem, Harald Bo says in his interesting writings: < Does a
day exist? In any case not a whole day because the beginning already disap-
peares before the end comes. The same is true for an hour and for a second.
The only time-interval which can exist is where the beginning and the end
are so close together that they are identical. That means a quantum zero-time
... We cannot put the last day on the table or somewhere else. We have no time in the physical sense. What we have is a psychological time >*".

("*Wie gesagt ist der Begriff der Zeit eigentlich etwas Unverständliches. Wir
sind es gewöhnt, zu denken, Zeit sei etwas, das man findet oder hat. Doch sie
hat eigentlich keine Existenz. Zeit ist etwas, das mit der einen oder anderen
Art von Uhr gemessen wird, werden die Physiker sagen. Aber was ist es,
was die Uhr mißt? Doch die Zeit!*

In seiner interessanten Schrift über das Problem sagt Harald Bo: <Existiert

ein Tag? Auf jeden Fall nicht der ganze Tag, weil der Beginn bereits ver-
schwunden ist, bevor der Schluß kommt. So ist es auch mit der Stunde und
so auch mit der Sekunde. Der einzige Zeitraum, der existieren kann, ist
der, wo der Beginn und der Schluß einander so nah kommen, daß sie zusam-
menfallen. Das bedeutet ein Quantum Nullzeit ... Wir können den gestrigen
Tag nicht auf den Tisch legen oder sonstwohin. In physikalischem Sinn ha-
ben wir keine Zeit. Was wir haben, ist eine psychische Zeit.>")

Final remarks

Now one gets the impression that the problems in connection with
space and time are becoming more and more complicated: Besides
Mach´s principle there is the *test-body effect* as an additional necessary
condition, and this effect reflects the fact that we can only observe distances and time-intervals in connection with material objects and
physical processes, respectively, but never with isolated coordinates
and times. However, we will show in the next sections that the situation does not become more complicated because of this effect; the opposite seems to be true: When we consider the test-body effect in the
space-time conception from the beginning, Mach´s principle is automatically satisfied. However, we have to give up the conception that
material objects are *embedded in* space-time; as we have outlined above, this conception is problematic anyway. This topic will be justified
in the next sections.

1.1.8 Can Space and Time be Considered as Metaphysical Elements?

Our space-time feeling is spontaneous, i.e., it appears in the perceptions of everyday life without the observer's intellectual involvement.
We feel that the things around us are embedded in space and time,

and in the following, we would like to call this space "everyday-space". To the points of everyday-space, the set of coordinates (x, y, z) is assigned.

The following is characteristic: *Everbody* has this kind of space-time feeling, i.e., *everybody knows* this space-time but noboby is able to determine empirically its basic elements x, y, z and t which must therefore be considered as *metaphysical elements*.

On the other hand, since *all* observers have this kind of space-time feeling we may state that space-time is in certain respects *objective* in character, and nobody would deprive somebody of this space-time feeling. In other words, space-time is *objectively real* and, on the other hand, *metaphysical* because its elements x, y, z, t are not accessible to empirical tests.

In connection with the facts discussed above, one hesitates somehow to consider space and time as metaphysical elements. This feeling is further strengthened when we consider usual metaphysical elements, i.e., quantities and statements which are also not accessible to empirical tests.

To these usual metaphysical elements belong, for example, fields which cannot be experienced because of their infinite number of values [4]. Also *"so intimate and highly assessed statements as both laws of thermodynamics or the assertion that matter consists of once and for all, i.e., absolutely given "ultimate constituents" (elementary particles)"*[4]. (*"so vertraute und hoch bewertete Aussagen wie die beiden Hauptsätze der Thermodynamik, oder die Behauptung, daß die Materie aus ein für alle mal, also absolut gegebenen "letzten Bausteinen", eben den Elementarteilchen, zusammengesetzt ist"*[4]). But these metaphysical quantities and statements are essentially different from space and time:
1. These are quantities which are not spontaneously accessible; they

are constructions of the human mind and, therefore, they are in certain respects not accessible to *all* observers, i.e., they are not objectively existent, as is the case for everyday-space.

2. Because of this feature we may possibly have to give them up at some time; this can be the case when a new physical world view with principally new concepts comes into play. It is however not conceivable that the spontaneous space-time feelings in everyday life, which are equally experienced by all observers, can be replaced at some day by another feeling. In other words, the usual metaphysical elements can in principle be eliminated, which is obviously not possible in the case of everyday-space.

Because of these facts, in particular because of the objective existence of space and time, we hesitate to put these quantities into one pot with the usual metaphysical elements. There is something different with space and time. But what? In the next sections we will develop a conception which permits the following view:

All *usual* metaphysical quantities and statements (fields, concept of the elementary particle, etc.) are used for the characterization of reality. This is different in the case of space and time because they do not primarily describe reality but they are tools in order to be able to describe or to observe reality. Then we have to consider space and time as "instruments" of observation. Then, the observation of space and time means to make an "observation of the observation". We then leave the level of reality which we normally observe. In this way we can understand on the one hand why space and time are objectively real and, on the other, why they are not accessible to empirical tests when we use the usual methods and not those which are adequate on a higher level of reality. Only from this point of view is the contradiction that space and time are *simultaneously objectively real* and *metaphysically* apparent.

We will see in the forthcoming sections that this conception is based on the view that material objects are *not* embedded in space-time which - as we have outlined above - in any case is a problematic view. Instead of this we will come to the conclusion that the world (all things) which we see in front of us is not reality itself but "only" a picture of reality.

1.2 Is the World Embedded in Space ?

1.2.1 The Colour Experiment

A reproducible, non-material phenomenon

If we look with good illumination at the little white dot situated within the coloured figure in Fig. 3 for approximately one minute and look thereafter at a white area, this figure appears on this white area, although it does not exist there. Furthermore, the figure does not have the original colours (Fig. 3), but they are complementary to the colours in Fig. 3.

It is relatively easy to explain this effect. In [17] we find:
"The phenomenon is due to a sign of fatigue of specific colour receptors of our retina. The impression "white" arises by the uniform excitation of all colour receptors. If we fix the presentation for an extended time, part of them will be fatigued, so afterwards the activity of those receptors which have not been strained predominates and, therefore, their value of colour normally passed on dominates temporarily the foreground in our experience."
("Das Phänomen ist auf einen Ermüdungseffekt spezifischer Farbrezeptoren in unserer Netzhaut zurückzuführen. Die Empfindung "weiß" entsteht durch eine gleichmäßige Erregung aller Farbrezeptoren. Wird ein Teil von ihnen durch längeres Fixieren der Vorlage erschöpft, so überwiegt anschließend die Aktivität der nicht beanspruchten Rezeptoren, wodurch in unserem Erleben der durch sie normalerweise übermittelte Farbwert vorübergehend in den Vordergrund tritt.")

The colour experiment is reproducible. Under normal circumstances

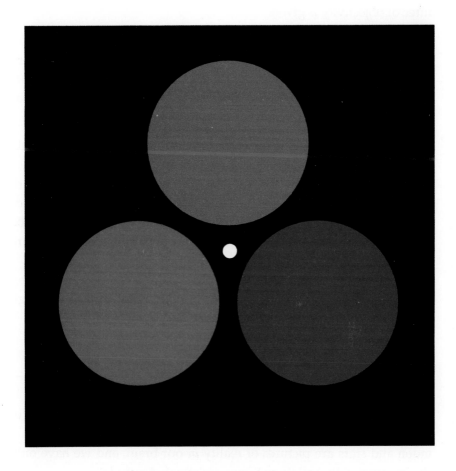

Figure 3. By means of this coloured figure on the black area we can perform reproducible experiments. Details are given in the text.

every person is able to experience this phenomenon, in so far the guarantee of objectivity is given.

We see the coloured figure *outside* on the white area. The explanation of the effect is however given in terms of colour receptors which are inside the brain. This means that the figure cannot be on an area *outside* but everything (white area, figure, etc.) is *inside* the head. We have only the impression that everything is situated *outside* in the space.

In conclusion, the colour experiment with Fig. 3 makes two points clear:
1. The image of what we have in front of us is actually inside the head. We have only the impression as if everything were situated in a space *outside*. The coloured figure which we see on the white area cannot be experienced by another observer, and no apparatus could measure these colours.
2. Under normal circumstances, the phenomenon can be perceived by every person, in so far the guarantee of objectivity is given although the phenomenon is inherently *non-material*.

It is only a feeling that an object is standing opposite the observer

The colour experiment makes the following clear:
Everything that we see is primarily *in* our head; it is not outside. Not only the coloured figure but also persons, cars, aeroplanes, the sun, moon and stars are pictures of reality *in* our brain, and we have only the impression that all these things are situated outside.

These conclusions are supported in [18]:
"We have devices in the cerebral cortex which - comparable with a television screen - produce "pictures" in our awareness from the nerve-excitations coming from the retina. It is characteristic of the sight-process that our awareness does not register the picture of a candle on the retina inside the

eye, but we have the impression that we are standing opposite the candle-light which is situated in the space outside, not standing on the head but upright. All observations are "projected" by our senses outwardly into space. We see "real objects" in front of us and around us. Within this act of perception, the eye, nerve-wire and the brain work together. To see without the brain is as impossible as to see without eyes."

("In der Hirnrinde besitzen wir Vorrichtungen, die - vergleichbar etwa mit einem Fernsehschirm - von der Netzhaut hergeleitete Nervenerregung in unserem Bewußtsein wieder als ein "Bild" aufleuchten lassen. Charakteristisch für den Sehakt ist dabei, daß unser Bewußtsein nicht das Bild einer Kerze auf der Netzhaut drinnen im Auge registriert, sondern in uns die Empfindung wachgerufen wird, einer Kerzenflamme gegenüber zu stehen, die sich - nicht auf dem Kopf, sondern aufrecht - außerhalb von uns draußen im Raum befindet. Unsere Sinne "projizieren" alle Wahrnehmungen nach außen in den Raum. Wir sehen "leibhaftige Gegenstände" vor uns und um uns herum. Bei dieser Leistung arbeiten Auge, Nervenleitung und Gehirn zusammen. Sehen ohne Gehirn ist ebenso unmöglich wie Sehen ohne Augen.")

In conclusion, this impression produced by the brain, that everything, which is situated outside in the space, represents a

picture of reality,

it is a picture which is localized in the head, i.e., what we see in front of us is not the actual reality outside. Information about reality outside flow via our senses into the body, and the brain forms a picture of reality; the situation is represented schematically in Fig. 4.

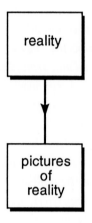

Figure 4. Information about reality outside flow via our senses into the brain, and the brain forms a picture of reality.

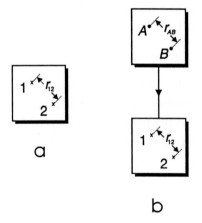

Figure 5. Explanations are given in the text.

1.2.2 Scientific Realism

Are the structures and characteristics in the picture identical with those of reality outside?

Most people assume automatically that the things in front of them is material reality itself, and those which are conscious of the fact that it is only a picture, normally assume as a matter of course that the structures and the other characteristics in the picture are identical with material reality outside. It is evident that there are no material objects in the picture but only geometrical structures, and that what we see in front of us is a picture of reality and not material reality itself. However, as already mentioned, it is assumed by most people that the structures in the picture are identical with those in the actual world outside.

For example, let us consider two material objects (for example, two planets) which appear in the picture in front of us as two geometrical positions 1 and 2 and these geometrical positions are symbolized by crosses in Fig. 5a; r_{12} is the distance between 1 and 2. We assume in the most natural way that the composition in the picture is identical with the situation in reality, i.e., we assume that in reality the geometrical positions in the picture are replaced by material objects (Fig. 5b) where the full points A and B represent the material objects in actual reality outside; the distance between the full points A and B is then given by $r_{AB} = r_{12}$ because the geometrical points (crosses) are merely replaced by the material objects. But is that realistic? In other words, is the conclusion from Fig. 5a and Fig. 5b verifiable or compatible with other characteristics in natural sciences? In the next sections we will show step by step that this is not the case, and this is supported by a large number of indications.

The picture-independent point of view

How can we verify that the situation in Fig. 5b actually exists? In other words, how can we verify that the compositions in the picture are identical with those in reality outside? We cannot because reality is only accessible to an observer on the basis of a picture, i.e., a picture-independent point of view is not conceivable.

The role of the equations of motion

Classical mechanics claims to describe the motion of material objects as it really takes place in actual reality. Newton´s mechanics is based on observations which the observer experiences in everyday life, i.e., these are experiences on the basis of pictures which are directly in front of the observer. Newton´s equations of motion describe material objects on their way through space and time. However, in the picture, not the *material objects* are moving relatively to each other, but the *geometrical positions* are moving. This fact is then unimportant for the equations of motion if we assume that the composition of actual reality is identical with the structure in the picture, i.e., when the scheme in Fig. 5 is valid.

Newton itself certainly assumed that we are "embedded in space" whereas the scheme in Fig. 5 indicates that it should be a "projection onto space", i.e., reality is projected onto space. That is a difference in principle which is however not relevant if the structure in the picture is identical with that in reality.

However, Fig. 5 reflects a scientific realism which, in our opinion, is questionable; such a conception would be, for the following reason, simply too naive:
Within the framework of Newton´s theory the masses and the the interactions between them are considered to be objectively real; also

the solutions of Newton's equations of motion, i.e., the trajectories of the masses, are considered to be objectively real because we observe with high precision the motion of a celestial body (for example, the moon) just as the equations of motion predict. In conclusion, the equations of motion with their elements and solutions reflect how reality is composed, and this is independent of the conception of whether the material objects are *embedded in space* or whether they are *projected onto space* (in accordance with Fig. 5).

Such a scientific scenario would involve that the masses solve, in their motion through space, incessantly differential equations (Newton's equations of motion). But such a realism seems ridiculous which is also reflected in the following remarks [19]:

"As Herschel ruminated long ago, particles moving in mutual gravitational interaction are, as we human investigators see it, forever solving differential equations which, if written out in full, might circle the earth."

If we use, however, Fig. 4 and Fig. 5 as a basis, the equations of motions refer primarily to the elements in the picture (as discussed above, we have in any case only pictures in front of us and not reality itself). But this does not mean that nature solves incessantly differential equations because the picture is a construction of the biological cognition apparatus just as the equations of motion are the result of a cognition process. If we use the equations of motion only for the description of the elements in the picture, the conception that the masses solve, in their motion through space, incessantly differential equations is avoided. Then, we have to conclude that Fig. 5 cannot be correct. This is because we have now to assume that there is in general no similarity between the structures and characteristics in the picture (Fig. 5a) and those in actual reality outside. How actual reality is constructed cannot in principle be said because - as already remarked above - there does not exist a picture-independent point of view for the observer.

Thus, instead of Fig. 5, we obtain Fig. 6.

Figure 6 is supposed to show that the world actually is *not* as science takes it to be, and that its furnishings are *not* as science envisages them to be [19]:

"Scientific realism is the doctrine that science describes the real world: that the world actually is as science takes it to be, and that its furnishings are as science envisages them to be. If we want to know about the existence and the nature of heavy water or quarks, of man-eating molluscs or a luminiferous aether, we are referred to the natural sciences for the answer. On this re-alistic construction of scientific theorizing, the theoretical terms of natural science refer to real physical entities and describe their attributes and compo-nents. For example, the "electron spin" of atomic physics refers to a be-haviourial characteristic of a real, albeit unobservable, object - an electron. According to this currently fashionable theory, the declarations of science are - or will eventually become - factually true generalizations about the actual behaviour of objects that exist in the world. Is this "convergent realism" a tenable position?

It is quite clear that it is not. There is clearly insufficient warrant for and little plausibility to the claim that the world indeed is as our science claims it to be - that we've got matters altogether right, so that our science *is cor-rect* science *and offers the definite "last word" on the issues. We really can-not reasonably suppose that science as it now stands affords the real truth as regards its creates-of-theory."*

This comment by Rescher shows, together with the above discussion in connection with the role of the equations of motion, that we have to assume that the scientific pictures have only limited similarity with what actually takes place in reality outside. Since, on the other hand, the scientific pictures are essentially based on our observations in eve-ryday life (in particular, they have to be compatible with them), we may also conclude that those things and processes which we have in

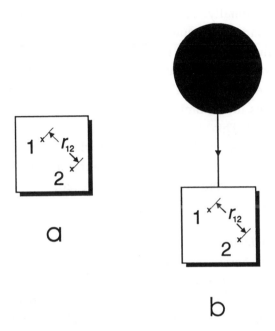

Figure 6. The full black circle symbolizes the actual reality. It cannot be observed directly by the observer since there does not exist a picture-independent point of view. Statements about actual reality can only be made by means of pictures.

in front of us in everyday life can also have only limited similarity with what actually takes place in reality outside. Let us deepen this discussion with the help of some examples.

1.2.3 Quality and Quantity

Inventions of the biological apparatus:
Light and colour, music, heat and cold, pain

There exist characteristics in the picture which cannot exist in reality outside. When we trace *quality* back to *quantity*, which is one of the main tasks in physics, the character of the phenomenon is changed in general fundamentally. For example, the quality "colour" is traced back to the quantity "frequency of oscillation" which is the physical quantity for the characterization of the electromagnetic wave. Because of their *objectivity* the physical quantity is considered to be objectively real but not the colour. It is the real something (electromagnetic wave) which produces in the brain of the observer a certain quality (for example, colour). This however means that in the picture of reality there occur features (for example, colour) which do not occur in actual reality outside. This is certainly the present opinion. Concerning this point, the following comment is of relevance [20]:

"The eminent cyberneticist Heinz Von Förster points out that the human mind does not perceive what is "there", but what it believes should be there. We are able to see because our retinas absorb light from the outside of the world and convey the signals to the brain. The same is true of all of our sensory receptors. However, our retinas don't see colour. They are "blind", as Förster puts in, to the quality of their stimulation and are responsive only to their quantity. He states, < This should not come as a surprise, for indeed "out there" there is no light and no colour, there are only electro-magnetic waves; "out there" there is no sound and no music, there are only periodic variations of air pressure; "out there" there is no heat and no cold, there are only moving molecules with more or less mean kinetic energy, and so on. Finally, for sure, "out there" there is no pain. Since the physical nature of the stimulus - its quality - is not encoded into nervous activity, the fundamental question arises as to how does our brain conjure up the tremendous varity of colourful world as we experience it any moment while awake, and sometimes in dreams while asleep >."

The question is whether *all* elements in the picture are inventions of of the biological apparatus as it is obviously the case for colour, light, music, etc. In other words, does the *picture of reality* consist completely of an ensemble of symbols which do not occur in this form in reality outside? This question contains a serious supposition which is however obvious and this is because of the above-discussed features of *quality* and *quantity* which we have explained in connection with colour.

This supposition is identical to what we have outlined above in the subsection "Role of the equations of motion".

In conclusion, our analysis leads to the following thesis:
The picture of reality contains admittedly concrete information about reality outside, but there is no similarity between reality and the corresponding picture.

For this thesis, we have already given good arguments and proofs, respectively, but it so far-reaching that we have to give more evidence.

1.2.4 Space and Time as Features of the Picture

Direct and picture-dependent observation

Figure 5 is based on the conception that the situation "outside" is exactly copied inside the observer. However, we have seen above in many different respects that this cannot be the case, not to mention the fact that the actual reality outside is in principle not accessible, and this is because the observer is not able to take a *picture-independent* point of view, i.e., there is no external point of view which would enable a *direct* observation of reality. Then, it becomes senseless to ask for the features of actual reality outside.

The conclusions concerning the situation in Fig. 5 must in particular also be valid for *space* and *time*. Even when we ignore the objection of the picture-independent point of view in connection with space and time, the following question arises: How could such an exact transmission from reality into the brain (e.g., in accordance with the situation in Fig. 5) take place at all? In Chapter 1, we have stated that the coordinates x, y, z and time t, i.e., the space-time points, are not accessible to empirical tests. But if the human body wants to copy, by means of its biological cognition apparatus, reality outside he necessarily has to experience reality outside, i.e., also the space-time points which are however not accessible to empirical tests, as we discussed in Subsec.1.1. How can information about space and time from reality outside enter the brain? It cannot because there does not exist a possibility for that. In other words, from this point of view space and time are *internal* elements, i.e., space and time are exclusively elements of the picture; it makes no sense to assume that they are also elements of reality outside; a coordination of the material objects and processes in actual reality on the basis on space and time does not exist. Instead of Fig. 5, we get the scheme in Fig. 6.

The fact is that the coordinates and time are elements of a fictitious net which the observer intellectually puts over the picture in front of him. That is all what we can say about the space-time elements because more far-reaching statements cannot be proved experimentally. Consequently, we are fixed concerning space and time. Therefore, only the statement that x, y, z and t are *picture-specific* elements is actually realistic because these elements can only appear in the picture.

This point of view is distinctly illustrated by an experiment with distorted glasses which we would like to discuss in the next section.

Experiment with distorted glasses

A conception of reality in the sense of Fig. 5 assumes a stringent correlation between the space inside the observer and the space in the reality outside. Changes in reality are transmitted *one to one* onto the picture; that is immediately clear, for example, in the case of motion. The cause for a phenomenon in the picture is an effect in reality outside, i.e., the effect in reality outside is primary, the phenomenon in the picture however is secondary. Only in this way we can understand Fig. 5.

If however changes occur in the picture although the situation in reality outside remains constant we can no longer state that we have a *one to one* representation in the sense of Fig. 5. Only such effects occur when we perform experiments with distorted glasses:
If one uses glasses which strongly distort the spatial situation, after a certain time, we see everything in the normal order, i.e., the spatial situation is the same as before without distorted glasses and space is again experienced as right-angled (euclidean). In other words, the distorting glasses are ignored after a certain time by the observer's cognition apparatus.

In conclusion, distorted glasses transform a space, which satisfies the axioms of euclidean geometry, into a space with non-euclidean geometry. Without changing the physical situation, we observe after a certain time the following strange effect: The cognition apparatus of the observer seems to transform space with non-euclidean geometry into usual space which satisfies the axioms of euclidean geometry. In other words, the cognition apparatus of the observer is able to influence the space and picture, respectively.

Space is changed although the physical system and the influences due to the outside world remain constant. We have stated above that the scheme in Fig. 5 can only be understood if an effect on reality outside (primary effect) produces a phenomenon in the picture (secondary effect). The experiment with distorting glasses shows however that this is not guaranteed. That means that we have to reject, strictly speaking, the scheme in Fig. 5; phenomena come into play which do not fit the scheme in Fig. 5.

The observer with distorting glasses, which however experiences after a certain time a picture which is free of distortions (i.e., the geometry again becomes euclidean after a certain time) has to conclude, by reversal of the optical beam path, that the geometry in the outside world is non-euclidean. What kind of geometry is there in the outside world, is it euclidean or non-euclidean?

The situation with glasses is epistemologically equivalent to the situation without glasses; the reason why one would give preference to the situation without glasses is a matter of habit, because evolution has developed just such a system.

However, if the evolution had developed an eye with the features of distorting glasses we would have the following situation: Observers with such eyes would assert that the geometry of the outside world is non-euclidean. If these people have glasses which rectify space, i.e., glasses which transmit the non-euclidean geometry into an euclidean one, we can expect that after a certain time the euclidean space-geometry will be transmitted into a non-euclidean space-geometry, as in the above case where the distorted space with non-euclidean geometry became a rectified space with euclidean geometry. This conclusion is justified by the following fact: *"What a man sees depends both upon what he looks on and also upon what his previous visual-conceptual experience has thought him to see."*[28]

In conclusion,different biological systems allocate to the world geometries which are in general different from each other. Which geometry can be considered as true? Since the biological systems are in principle equivalent, we have to conclude that both geometries must be true. This contradiction is dissolved when we assume that no geometry is realized in actual reality outside, but that space and time with their different geometries exclusively occur in the picture, i.e., space and time must be considered as *picture-specific* elements.

Thus, also from this point of view we obtain the scheme which is given in Fig. 6:

Actual reality is not accessible in a direct way because no picture-independent point of view is conceivable. There will also be no space and time in actual reality. Pictures can be formed in principle on the basis of different space-time structures. In general, there will be no similarity between the structures and characteristics in reality outside and the corresponding picture.

1.2.5 The Chick Experiment

The thesis which we have just formulated is daring. In particular, it means that different biological systems should form pictures of the *same* reality which are in general different from each other, and there can even be no or almost little similarity between these pictures. Is there evidence of that? Yes, there is. Let us briefly discuss this point by means of an interesting experiment which has been performed within the framework of behavioural research.

One of the natural enemies of the turkey is the weasel. It is its deadly enemy. When a weasel approaches the nest of a turkey, it protects its chicks and defends them against the attacker with violent pecks.

Using this system (consisting of a turkey, its chicks and a weasel) *Wolfgang Schleidt* performed some interesting experiments. These and similar investigations could be of such importance as certain key experiments in physics which have fundamentally changed the scientific world view.

What kind of experiments have been performed by Schleidt? He has not used expensive and complicated experimental devices as, for example, are customary and necessary in elementary particle physics but he worked with almost-everyday methods.

One knows that a turkey which sits on its just-hatched chicks attacks everything which approaches its nest. This is of course not true in the case of any of its own chicks which has - for any reason - left the nest. In order to protect the chick it will steer back the squeaking little animal with calming calls into the nest. All that seems to be more than a matter of course; the turkey really shows almost human behaviour.

The fact is however that the perception apparatus of turkeys must be quite different from that of the human observer. This can be demonstrated by means of two simple experiments.

1. Schleidt blocked the ears of the turkey so that it could not hear anything. After a certain calming time one of her chicks approached the nest and a serious disaster happened: Without hesitating the turkey strongly pecked the chick with its beak until it was dead.

The turkey saw its chick approaching but did not identify it. Everything that is "unknown" and that approaches its nest will be fought.

2. Schleidt implanted into the body of a stuffed weasel, enemy of the turkey, a little loudspeaker which emitted the sound of a squeaking little chick. By means of a hidden device he moved the stuffed weasel up to the nest. Also in this case something happened which was quite unexpected: The turkey saw the weasel come but did not identify it; after some hesitation, it even protected its enemy.

These dramatic and unexpected shocking results lead to the conclusion that the turkey must optically experience the world quite differently than we do although the eyes of the turkey are quite similar to ours. There is obviously no similarity between what the turkey experiences and what the human observer sees in the same situation!

Not only *one* turkey developed such a world view but *all*. In other words, the experiments are *reproducible*, i.e., it is scientifically acceptable.

In summary, we can say that *the pictures which are formed by different biological systems from reality outside are in general different from each other*. In particular, the experiments by Schleidt support impressively that our thesis made in the Subsec. 1.2.1-1.2.4 are obviously realistic.

1.3 Pictures of Reality

Let us briefly summarize the main points which we have developed in Subsec. 1.2:

1. We do not have actual reality outside with their "hard objects" immediately in front of us but "only" *pictures* of it; in those pictures the "hard objects" appear as *geometrical positions*.
2. Space and time are *picture-specific* elements; they occur in reality outside.
3. The structures and characteristics of reality are not accessible in a direct way because obviously no picture-independent point of view is conceivable.
4. There is in general no similarity (no one-to-one correspondence) between the structures and characteristics in the picture and those in actual reality outside.

1.3.1 Strategy of Nature

It is conceivable that conceptions of the world, which are different from each other, are *equivalent* concerning their information and efficiency. This is not a contradiction because world views of different biological systems, which cannot be compared with each other, can be equivalent.

In accordance with the above discussion, higher developed organisms (for example, an organism which could be descended from the turkey) could have developed epistemologically equivalent conceptions in comparison to man, although the conception of one system is different from the other.

Those world views, developed by different biological systems, are not *wrong* but, on the other hand, are also not *true* in the sense of *a precise reproduction* in the way that a photography represents a one-to-one copy of the world. That the features in the pictures cannot be wrong and, on the other hand, not true in the sense of a precise reproduction, can already be extracted from the *strategy* (principles of evolution) on which nature is based.

An important basic principle: As little outside world as possible

The perception of *true* reality in the sense of a precise reproduction implies that with growing fine structures in the picture, increasing information of actual reality outside is needed. Then, the evolution has developed sense organs with the property to transmit as much information from reality as possible. But the opposite is correct: The strategy of nature is to take up as little information from the outside world as possible. Reality outside is not assessed by "true" and "untrue" but by "favourable towards life" and "hostile towards life". Concerning this point, *Hoimar v. Ditfurth* stated the following [17]:

"No doubt, the rule < As little outside world as possible, only as much as is absolutely necessary>, is apparent in evolution. It is valid for all descendants of the primeval cell and therefore also for ourselves. Without doubt, the horizon of the properties of the tangible environment has been extended more and more in the course of time. But in principle only those qualities of the outside world are accessible to our perception apparatus which, in the meantime, we need as living organisms in our stage of development. Also our brain has evolved not as an organ to understand the world but an organ to survive."

("Kein Zweifel, die Maxime "So wenig Außenwelt wie möglich, nur so viel,

wie unbedingt notwendig", hat der Entwicklung ihren Stempel aufgeprägt. Sie gilt für alle Nachkommen der Urzelle und damit auch für uns selbst. Der Horizont der faßbaren Umwelteigenschaften ist im Laufe der Zeit ohne Zweifel immer weiter geworden. Grundsätzlich aber sind auch unserem Wahrnehmungsapparat nur die Qualitäten der Außenwelt zugänglich, die wir inzwischen auf unserer erreichten Entwicklungsstufe als lebende Organismen benötigen. Auch unser Gehirn ist ursprünglich kein Organ zum Verstehen der Welt, sondern ein Organ zum Überleben.")

The principle "as little outside world as possible" can be understood by means of the idea of evolution which is briefly discussed in the next section.

1.3.2 The Principles of Evolution

The *principles of evolution*, i.e., the phylogenetic development from simple, primeval forms to highly developed organisms, can be considered as the key for the perception of reality of biological systems (for example, man and turkey). It is the theory of evolution by natural selection which is generally accepted in the meantime; its foundations have been created by *Charles Darwin* (1809-1882) more then one hundred years ago. Since then it has been modified and developed further by geneticists.

Evolution by natural selection is a two-step process:

Step 1
By recombination, mutation, etc., genetic variants are produced at random. Populations with thousands or millions of independent individuals arise.

Step 2
Some of these independent individuals will have genes which enable

them to manage the predominating situation due to the enviroment (climate, competition, enemies) better than others. They thus have a larger chance of survival than others; they will have, in the statistical avarage, more descendants than other members of the population. Natural selection takes place in favour of those organisms whose genes have adapted to better cope with the environment.

The number of examples that biological systems have developed in accordance with these criteria is overwhelming. In connection with our discussion, it is important also to mention that the *pictures of reality* formed by the different biologigal systems (for example, man and turkey) must also be characterized by this *species-preserving appropriateness*.

Not cognition but the differentiation between "favourable towards survival" and "hostile towards survival"

In conclusion, in nature cognition does not play an important role but the differentiation between "favourable towards survival" and "hostile towards survival", at least in the early phases of evolution. In this connection, it is important that one is able to recognize and to assess the earliest possible changes in the environment within the framework of a consistent *picture* (world view). The possibility of a fallacy has to be excluded as far as possible; there may be no doubts concerning the particular status of the environment. For this purpose, the *picture of reality* designed by the individual has to be correct but it may only contain, for economic reasons, information which is absolutely necessary for survival; everything else is unnecessary; it encumbers and is therefore hostile towards survival. The picture of reality does not have to be *complete* and *true* (in the sense of a precise reproduction) but *restricted* and *reliable*, at least during the early phylogenetical phase. These are the criteria which guarantee optimal chances of survival.

1.3.3 How is the Picture Formed?

Reality and picture, cinema and cinema ticket

In summary, from the point of view of evolution we may state that the impressions in front of us are not *precise reproductions* of reality but merely appropriate *pictures* of it, which are formed by the individual from certain information from the outside world.

These pictures are incomplete; they are correct but not necessarily true. The last point follows directly from the fact that different organisms, as, for example, man and turkey, form pictures which are distinct from each other. None of them can be wrong because the concerned species would inevitably die out in the case where it had a wrong picture of reality.

All these facts can be extracted from the principles of evolution. After that, it is of primary importance that we can reliably distinguish between "favourable towards survival" and "hostile towards survival" and not to form a true world view. An individual registers situations in the environment by a certain pattern, which are tailor-made for the particular needs of the species and which are relieved of the condition to be a precise reproduction.

So, for example, in order to find a certain place in a cinema, it is not necessary that a visitor gets at the pay desk a small but true model of the cinema, i.e., a precise reproduction of the cinema, which is reduced in size; a simple cinema ticket with the essential information is more appropriate. In this respect, the cinema ticket in Fig. 8 is the picture of the cinema; similarly, Fig. 7 is the picture of the chick - from man's point of view. We, however, do not know anything about the picture of the chick from the point of view of the turkey.

Figure 7. The figure shows a chick. Everybody would identify the object with a chick. A turkey, which could even be the mother of the chick, would however not identify the object in the figure with a chick.

ODEON CINEMA	movie:	**Casablanca**
	date:	**June 5, 1963**
	time:	**7.30 p.m.**
	row:	**7**
	seat:	**13**

Figure 8. The pictures which are formed by an individual of reality outside are incomplete; they are correct but not true in the sense of a precise reproduction; they are primarily useful.

So, for example, in order to find a certain place in a cinema, it is not necessary that a visitor gets at the pay desk a small but true model of the cinema, i.e., a precise reproduction of the cinema, which is reduced in size; a simple cinema ticket with the essential information as to where to go is more appropriate. In this respect the cinema ticket in this figure is the picture of the cinema; similarly, Fig. 7 is the picture of the chick - from man´s point of view. We, however, do not know anything about the picture of the chick from the point of view of the turkey.

Basic reality: Exists independently of biological systems

That also means that we do not know how reality is actually constructed, i.e., how it is constructed independently of the observer's perception apparatus; it is probable that we may never know because we are
at first caught by our own system. We have mentioned above several
times, that a picture-independent point of view is not conceivable.
Therefore, the *basic reality* remains hidden in principle.

1.3.4 Compatibility

Man and other biological systems have developed so that they are able
to *consistently* and *conclusively* solve specific problems. This is valid
for the conscious cognition apparatus as well as for the unconscious
one, and of course for the anatomical set up.

The ability to survive is based on the possibility to form conscious and
unconscious pictures of reality which are tailor-made to species-preserving principles. Any picture consists of certain elements which
have to be *compatible* with each other. In a certain sense, the situation
is comparable with a linguistic pattern (see also Fig. 9).

On the other hand, the elements of pictures of one species are in general not compatible with those of other species; in other words, the elements of pictures of different species are in general not interchangeable. If one tries it nevertheless, difficulties can appear as in the case of
the chick-experiment by W. Schleidt; he tried to project the elements of
pictures, which are characteristic of human behaviour, onto the picture of the turkey and the result was a disaster (see also Subsec.
1.2.5).

In the beginning was the Word.

Вначале было Слово.

Figure 9. A certain fact can be expressed linguistically in many but equivalent ways. For example, the fact "In the beginning was the Word" formulated in English cannot be recognized if we express, for example, the same fact in Russian. The symbols of the Russian language are quite different from those of the English language. Not only the symbols, but also rules after which the symbols are connected (grammar), are different from each other. Each language, i.e., English or Russian, forms a closed consistent system. However, a Russian cannot understand an English sentence if he does not speak English and, on the other hand, an Englishman cannot understand the Russian sentence if he does not speak Russian.

The same should be true for man and turkey. Both form pictures of the same chick which are obviously quite different from each other; a turkey cannot do anything with the chick pictured by man and, on the other hand, man probably cannot do anything with the chick pictured by the perception apparatus of the turkey. As we have stated in the text, the picture of a species does not have to be true in the sense of a precise reproduction but "only" useful.

Petrol into an electric car?

For comparison, it is not possible to drive an electric car with petrol on the one hand and, to drive a car with a petrol engine with the battery of an electric car on the other. Petrol is not needed for the electric car and, a battery is not appropriate for the engine operating on petrol. That is more than a matter of course but is rather mysterious in the case of the chick experiment. However, both cases are arranged according to similar principles.

1.3.5 Objectivity

What about the *objectivity* of experiences? In particular, what can we say obout objectivity when the results of the chick experiment are taken into account?

Equivalence of all observers

In connection with scientific observations, we talk of an *objective fact*, when the observer is interchangeable. This is one of the basic principles in science. The results of a measurement made by one observer should be identical with those of any other observer. Then, the object of the observation is considered as an *objective fact*. In other words, observers which are independent of each other agree on certain facts of the outside world; only such kind of information is recognized in physics and the equivalence of *all* observers represents an important basic principle.

Not only in the laboratory but also in everyday life (in particular, in connection with the identification of shapes) objective performances can be achieved without conscious activities. If one shows Fig. 7 to any person and asks what he sees, "A chick" will always be the answer; one were surprised if anyone would see a flower or a pot instead of a chick.

But, in a certain sense, the turkey also is an observer and the experiment in Subsec. 1.2.5 showed that it was not able to identify the chick optically, although it was able to do that in the presence of sound. We have to interpret the experiments by W. Schleidt in this way. As already mentioned above, the conception of the chick pictured by the turkey is also *not wrong*, it is only different from that in Fig. 7. That is the case for *all* turkies.

After the criteria about objectivity we can of course introduce the concept of objectivity into the "world" of the turkey. Then, the conception of a chick or any other fact must be identical for all turkies. How can we prove that? Very simply because we only have to show that all turkies react to the chick in the same manner. This is obviously the case and, therefore, we may state also that the chick is in the "world" of the turkies.

In summary, only those facts can be considered as *objective* which are experienced in the same manner and this is obviously satisfied for "the observer which belongs to the same species", i.e., the concept of objectivity is only applicable if the observers are "of the same kind". Since turkies and man do not belong to the same species, we may state that the two objectivities are not identical. "Objective" only means that "observer of the same kind" come in the assessment of a certain situation to the same result. "Objective" does not mean that a certain fact actually exists and takes place, respectively, in (*basic*) *reality* in the form experienced by an observer (turkey or man). In principle, we do not know how a certain fact (for example, a chick) is fixed in basic reality since a picture-independent point of view does not seem to exist.

1.4 Levels of Description, Levels of Reality

1.4.1 Nothing Observable Exists Independent of the Observer

"There are only few things of which the scientist is more convinced than of the real character of his experiments and the objectivity of his statements about them."

(*"Es gibt nur wenige Dinge, von denen ein Naturwissenschaftler mehr überzeugt wäre als die Realitätsbezogenheit seines Experiments und der Objektivität seiner Aussagen darüber."*)

This statement is given in [21], and it decribes the present situation correctly. Here, two concepts are in focus, *reality* and *objectivity*. Concerning these two concepts, let us briefly repeat the results which we have found in the preceding sections.

Different pictures of the same object

We do not know the *basic reality* (actual reality outside) and, in principle, we cannot make any statement about it. Our cognition apparatus can "only" form *pictures* of reality (for, example, a chick). This is also the case for the cognition apparatus of other biological systems (for example, a turkey).

It turns out that the cognition apparatus of different biological systems, as man and turkey, form different pictures from the *same* object of basic reality. We have to assume that there is no or almost no similarity between these pictures. Remember, the turkey vigorously pecked the chick with its beak until it was dead, although it could see the chick.

The pictures do not have to be *true* in the sense of a precise reproduction but "only" *conclusive* and their elements must be *compatible*. Petrol is not appropriate to the electric car and, a battery is not compatible with the engine running on petrol.

We have outlined that the connection between *picture* and *reality* should be similar to that of *cinema ticket* and *cinema*.

Furthermore, we have stated above that only those facts can be considered as *objective* which are experienced in the same manner and this is obviously satisfied for an "observer which belongs to the same species". Since turkey and man are not of the same kind we may state that the two objectivities are not identical. "Objective" does not mean that a certain fact actually exists in *basic reality* in the form experienced by an observer (turkey or man).

Pictures of the same reality on different levels

In nature, cognition does not play the important role but the differentiation between "favourable towards life" and "hostile towards life", at least in the early phases of evolution. Species have obviously developed on the basis of this principle, and the individuals were inventive to master difficult situations in their lives.

In the case of man, the improvement of specific situations in life requires in general a *conscious action*; here, conscious means to have assessed and understood a certain situation within reality. Then, we can try to produce by a conscious action a new improved situation - a new improved picture of the *same* reality, i.e., a picture of reality on another level. In this way we come to the concept of *level of reality*.

1.4.2 Basic Reality and Levels of Reality

Due to the statements, which we have made up to now, the following conception is suggested:

Basic reality, i.e., reality which exists independently of the observer, is principally not accessible in a *direct* way. But it is observable or describable by means of pictures on different levels, i.e., *levels of reality*.

Hierarchical structure in accordance with the degree of generality

According to what principle are these levels of reality arranged? The levels can be arranged vertically in accordance with the *degree of generality* (see also Fig.10), where the level with a higher degree of generality is arranged above that with a lower degree of generality. In the following we will call this principle "principle of level-analysis".

These levels should be considered as *equivalent* if the pictures belonging to them instruct the individual to *equivalent actions*, where the pictures may be quite different from each other.

This does not mean at first that such pictures must have the same content of information, because the concept of information is obviously more connected with the features "true" and "untrue" than with "favourable towards survival" and "hostile towards survival" which however are of primary importance, as we have discussed above.

An upper limit for the number of levels cannot be given. In principle, a certain level can be produced from another level by any small change. Then, both levels are distinct from each other but are close together, i.e., the distance between them can be very small. On the other hand, there must be no connection between different levels of reality because different pictures can be incommensurable, as in the case of the chick experiment.

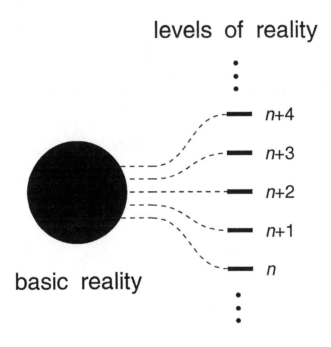

Figure 10. In principle, we cannot make any statement about *basic reality* (actual reality outside) which is symbolized in the figure by the full circle. But we can observe or describe aspects of it within the framework of *levels of reality* which are vertically arranged in accordance with the degree of generality (*principle of level-analysis*). The levels up to n-1 and those above n+4 are not quoted; n is any number.

"Additions" to the levels of reality

We have emphasized above that a new reality can be produced by conscious action which can be - according to the success - an improvement on the old one. Conscious action implies a process of thinking, i.e., "something" flows in from the cognition apparatus, we shall call it "addition". In conclusion, with the help of "additions" we can proceed from one level (level n in Fig.11) to another (level $n+1$ in Fig. 11).

"Addition" not only means to add details to a frame, which is already conceptually fixed, but the transition from one level to another which is more general than the old one, where the conceptually fixed frame of one level has little to do with that of the other level.

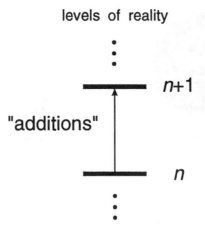

Figure 11. No species can make statements about basic reality. But it is observable or describable by means of pictures on different levels, i.e., levels of reality. In order to be able to master difficult situations in life, the creation of a new reality is often necessary; this can be done with the help of "additions", and we can come from one level (level n) to another (level n+1). In the figure, the levels n and n+1 are arbitrarily chosen from a set of levels of reality.

Examples

In Subsec. 1.2.3 we have discussed the usual conception of reality which does not consider the *principle of reality-analysis.*

This becomes particularly plain in connection with the formulation used in Subsec. 1.2.3:
"For example, the quality "colour" is traced back to the quantity "frequency of oscillation" which is the physical quantity for the characterization of the electromagnetic wave. Because of their *objectivity,* the physical quantiy is considered to be objectively real but not the colour. It is the real something (electromagnetic wave) which produces in the brain of the observer a certain quality (for example, colour)."

In contrast with that, i.e., to the *usual* point of view, we have to use instead, on the basis of the *principle of reality-analysis,* the following arguments:
Because of their objectivity, certain quantities in physics are considered to be objectively real. This is however not in accordance with the *principle of level-analysis* because we cannot say anything about the outside world, which we have called *basic reality.* For this reason neither "electromagnetic waves" nor the phenomenon "colour" belongs to the *basic reality;* both conceptions are equally pictures of one and the same phenomenon in the *basic reality.* The picture "electromagnetic waves" is associated with a level of reality which is in the hierarchy above that level which is associated for the picture "colour". This is because we have arranged the *levels of reality* in accordance with the degree of generality (Fig. 10). The picture "electromagnetic waves" is given by a set of equations and a set of equations are more general than a directly-experienced colour, which only represents *one* specific phenomenon, whereas the equations are valid for *all* possible electromagnetic phenomena.

The same can be said in connection with Newton´s equations of motion. These equations involve all possible orbits and not only those which we are just observing with or without measuring instruments. The "observation" of the equations of motion however means that we have "registered" at once *all* possible orbits.

1.4.3 In Any Case Pictures

All the things in front of us are pictures, also in the case of the so-called "hard" objects which are reflected by our five senses in everyday life. The "hard" objects are therefore features on the
level of everyday life
which we can also call *material* level of reality. Trees, stones, etc., are structures in pictures which do not exist in the same way, i.e., as precise reproductions, in *basic reality*. We have pointed out above that the connection between the *picture* (trees, stones, etc.) and the *basic reality* should be similar to that of *cinema ticket* and *cinema*.

Trees, stones, etc., are models which are designed in space and time by the cognition apparatus *without* our conscious action. In other words, the cognition apparatus "models", i.e., it produces theoretical pictures of trees, stones, etc. Since within such modelling, everything is managed without conscious action of the observer, we may call this process "unconscious thinking".

As is well known, the sensory perceptions can be extended by means of measuring instruments, i.e., we can penetrate deeper into reality by means of measuring instruments. Let us denote the level of everyday life which is extended by measuring instruments as
level of measuring instruments.

The level of measuring instruments is therefore by definition above the level of everyday life.

As we have already pointed out above, the scientific laws are also pictures of the same reality, which are however established on other, i.e., non-material, levels of reality; these pictures cannot be registered by our sense organs.

The scientific laws (as, for example, Newton´s equations of motion) are established on levels of reality which are in the hierarchy above the level of everyday life and also above the level of measuring instruments. This is because the experimental result can only be ordered on the basis of a theoretical picture, which is more general than the specific experimental result. We have already discussed this point in Subsec. 1.3.4 in connection with the equations of motion and electromagnetic waves. *Where* the idea comes from, on which the theoretical picture is based, remains an open question for the time being.

Nevertheless, one can object, despite all comfortable explanations and analogies, there is an irreconcilable difference: The "hard" objects belong to the *outside world* whereas the scientific laws are "things" or pictures *inside the brain*. However, we have discussed extensively above that this conception is wrong; everything that we know about "hard" objects are without exception pictures inside the brain. We cannot know directly what really takes place in *basic reality*; we cannot even guess it.

One of course never can observe in *basic reality* the effect of an action made by an observer, but only its effect on pictures of certain *levels of reality*. Clearly, it will be so that the person in action who wants to change a certain situation will use a picture of a certain level of reality. For example, if the person in action wants to move a chair from position A to position B, he controls his actions on the basis of the level of everyday life where the chair and the mortal body of the observer are arranged.

All "things" are similar in character

As mentioned, everything ("hard" objects and the scientific laws) are "things" and pictures, respectively, *in our head*. Because of this feature all "things", i.e., "hard" objects and the scientific laws, should be considered as *similar in character*. That we experience these "things" differently is due to the nature of the levels of reality, since to each level belong specific aspects.

1.4.4 To Objectify

We know how the picture of "hard" objects come into being: Our sense organs take up stimuli from *basic reality* which will be modelled to a picture by the cognition apparatus of the observer. Because the "things" are similar in character, the following question arises: Are scientific laws modelled by an analogous procedure? Here, it is sufficient to answer the question in an introductory way; the purpose of the following argumentation is to prepare the ground for a more solid discussion.

As the squeezing out of grapes?

Where do the scientific laws develop? *Karl Popper* in a discussion with *Franz Kreuzer* [22] made clear that the scientific laws cannot only be the result of direct experiences or measurements. The following remarks are instructive [22]:

"To grope in all directions. I do not favour that picture of science to gather and gather observations and from that the laws are distilled, so, as Bacon said, as wine is pressed from grapes. Within this conception, the grapes are

the observations and these grapes will be pressed and we will obtain wine, i.e., the generalisation, the theory. This conception is utterly wrong. It mechanizes the creative act of human thinking and inventing. To make that point clear is most important to me. Science proceeds in another way, namely to check ideas and world views. Science derives from myths. This can be recognized by the early scientists, namely the early greek, presocratic philosophers which were still strongly influenced by myths."

("Tasten nach allen Richtungen. Nicht also jenes Bild der Wissenschaft, daß man Beobachtungen sammelt und sammelt und daraus, wie Bacon gesagt hat, wie aus Trauben den Wein keltert. Die Trauben, die Beeren sind nach dieser Vorstellung die Beobachtungen, und diese Beeren werden ausgepreßt, und daraus kommt der Wein, das heißt also: die Verallgemeinerung, die Theorie. Dieses Bild ist grundfalsch. Es mechanisiert den schöpferischen Akt des menschlischen Denkens und Erfindens. Das ist, was mir das Wichtigste war. Die Wissenschaft geht ganz anders vor, nämlich sie geht so vor, daß sie Ideen, Weltbilder überprüft. Die Wissenschaft stammt vom Mythos. Man sieht es sehr deutlich bei den frühen Wissenschaftlern, nämlich den frühen griechischen, den vorsokratischen Philosophen, die noch sehr stark von der Mythenbildung beeinflußt sind.")

"To grope in all directions" means to check a physical world view via a dialogue with nature on a certain *level of reality;* in this way a world view can be improved and also be rejected if necessary. The theory should be reflected in many and, as far as possible, in all physical situations, and that has to be verified. Such a process can be called "objectivation by thinking".

A theory is useful when many variations of thinking and different experimental configurations *constantly* support this theory.

What about the "hard" objects which we have discussed in connection with the physical laws of theories? In [23], *Konrad Lorenz* (1903-1991) convincingly showed that the "hard" objects - as the scientific laws -

are groped by objectivation. This process reflects in the case of "hard" objects *unconscious* actions on the basis of physiological mechanisms, the so-called *constancy phenomena*. More details are given in the next section.

1.4.5 Constancy Phenomena

Objectivation in an unconscious, non-intellectual way (experience)

The recognition of certain situations in the environment requires *reliability* (reproducible occurrence). How has evolution realized that? It has caused the organisms to develop physiological mechanisms in the evolutionary processes; these mechanisms register automatically and objectively those signals from the environment which are relevant for the biological system. In this connection, the following is of interest: Of all the information about the environment only those which are relevant for the system are registered. In this connection, it is most important, that an object, observed by the individual, is - even in the case of large variations in the environment - unambiguously recognizable. In order to guarantee that, the evolution has developed so-called *constancy phenomena* whose *objectivation* performance is managed by the complex physiological apparatus. *Konrad Lorenz* stated [23]:

"Of special interest to the scientist striving for objectivation is the study of those perceptual functions which convey to us the experience of qualities constantly inherent in certain things in our environment. If, of course, we perceive a certain object (say a sheet of paper) as "white", even when different coloured lights, reflecting different wavelengths, are thrown on it, this so-called constancy phenomenon is achieved by the function of a highly complex physiological apparatus which computes, from the colour of the illumination and the colour reflected, the object's constantly inherent property which we call colour.

Other neural mechanisms enable us to see that an object which we observe from various sides retains one and the same shape even though the image on our retina assumes a great variety of forms. Other mechanisms make it possible for us to apprehend that an object we observe from various distances remains the same size, although the size of the retinal image decreases with distance."

(*"Von besonderem Interesse für den nach Objektivität strebenden Forscher sind jene Leistungen unserer Wahrnehmung, die uns das Erleben jener Qualitäten vermitteln, die gewissen Umweltgegebenheiten konstant anhaften. Wenn wir einen bestimmten Gegenstand, etwa ein Blatt Papier, in den verschiedensten Beleuchtungen in derselben Farbe "weiß" sehen, wobei die von ihm reflektierten Wellenlängen je nach Farbe des einfallenden Lichts recht verschieden sein können, so beruht dies auf die Funktion eines sehr komplizierten physiologischen Apparates, der aus Beleuchtungsfarbe und reflektierter Farbe eine dem Objekt konstant anhaftende Eigenschaft errechnet, die wir schlicht als die Farbe des Gegenstandes bezeichnen.*

Andere neurale Mechanismen ermöglichen es uns, die räumliche Form eines Gegenstandes bei Betrachtung von verschiedenen Seiten her als dieselbe wahrzunehmen, obwohl das auf unserer Netzhaut entworfene Bild sehr verschiedene Formen annimmt. Wieder andere Mechanismen setzen uns in den Stand, die Größe eines Objektes aus verschiedenen Entfernungen als gleich zu empfinden, obwohl die Ausdehnung des Netzhautbildes in jedem Fall eine andere ist usw. usf.")

In conclusion, the constancy phenomena allow the objectivation in an unconscious, non-intellectual way by the physiological apparatus; those processes and objects which are relevant for survival appear *directly* and *reliably* in front of us, i.e., without conscious action of the indi-vidual. It seems as if there is an independent reality, i.e., independent of the individual. Because of the *constancy* of the phenomenon, the experienced reality becomes *concrete* in character. Furthermore, this reality appears as a *complete* world since all unnecessary information about the environment have been eliminated.

In this conection the following remarks by *Konrad Lorenz* are instructive [23]:

"What causes us to believe in the reality of things is in the last analysis the constancy with which certain external impressions recur in our experience, always simultaneously and always in the same pattern, irrespective of variations in external conditions or in our psychological disposition."

(*"Was uns an die Wirklichkeit der Dinge glauben läßt, ist letzten Endes sicherlich die Konstanz, mit der gewisse äußere Einwirkungen immer gleichzeitig und in derselben gesetzmäßigen Verbindung miteinander in unserem Erleben auftauchen, allen Veränderungen der Wahrnehmungsbedingungen und der inneren Zustände unseres Ichs zum Trotze."*)

At this point we would like to emphasize once more that it is not of primary importance to have a *true* picture of reality in the sense of a precise reproduction. Those pictures are primarily of use to help the individual to successfully cope with life.

The registration of things (and processes) without conscious action as well as their reliable identification lets them appear to be *independent* and *concrete* in character. These are those entities which we have called above "hard" objects. The level of reality on which the "hard" objects are established is - corresponding to its elements - the material level. But we have always to consider that we identify in any case the elements in the picture with the "hard" objects since the outside world (*basic reality*) is principally not accessible.

In conclusion, those things which are directly registered (i.e., without conscious action) by our sense organs in connection with constancy phenomena are called *matter*.

The immediate execution and "visualisation" which was essential for survival at least in the phylogenetically early phases demand such concrete terms. However, as already mentioned, such concrete pictures on the material level are - in accordance with the principles of evolution - obviously only incomplete representations of reality: At least in the phylogenetically early phases all those things have been ignored, which were not of relevance for the naked survival of the individual; *outside information without any specific relevance are not registered by the sense organs and this is a result of evolution.*

Objectivation in a conscious, intellectual way (thinking)

The capability of thinking was developed during the later phase of evolution. The process of thinking makes it possible to extend the individual's knowledge about reality; connections can be analyzed and used to improve the conditions of life.

With the occurrence of mind, individual experience became possible, which was not possible without it. Individuals without a conscious mind handle certain situations in their life not with their own experiences, but with behavioural patterns which were developed by the members of thousands of generations of their own species. Such behavioural patterns however become without value or even harmful when the environment has changed crucially. We find a lot of examples for that in nature; species have become extinct due to changes in the conditions of the environment. By the process of thinking, we can adapt our behaviour due to our *own* experiences; the individual can now react to quick changes in the environment.

Here, thinking means objectivation in a *conscious, intellectual* way which we have already briefly discussed in Subsec. 1.4.4. In connection with the constancy phenomena, we have also recognized that on the level of everyday life, where the "hard" objects are established, an objectivation takes place. It is however objectivation in an unconscious, non-intellectual way by physiological processes.

In other words, the "hard" objects as well as scientific laws (world views) are registered in the same manner, namely by *objectivation*. Therefore, we may consider the conscious (intellectual) and the unconscious (non-intellectual) registration of facts as *analogous* processes.

This conclusion is supported by results which have been found within the frame of behavioural research. Konrad Lorenz remarks [23]:

"The physiological functions underlying these constancy phenomena are of the greatest interest in the context of theory of knowledge because they are exactly parallel to the process of deliberate, rational objectivation referred to above."

(*"All die physiologischen Leistungen, auf denen die Konstanzphänomene beruhen, sind erkenntnistheoretisch deshalb von so großem Interesse, weil sie der schon besprochenen Leistung der bewußten, verstandesmäßigen Objektivierung streng analog sind."*)

Remark:
By means of a scientific law, theoretical and experimental configurations are made conscious (isolated) by thinking, i.e., all these configurations are compatible with the "world" of scientific law. In other words, many variations of thinking and different experimental configurations reflect *constantly* this scientific law. Processes of thinking which carry out such selections can be considered as a *constancy phenomenon*.

1.4.6 Inside World and Outside World

Conscious and unconscious objectivations are analogous processes

Everything is located in the head, not only the products of fantasy and scientific laws, but also those which we have to understand as "hard" objects, and this is because we do not have the "hard" objects in front of us but "only" their pictures.

Then we can make the following statements about the concepts "inside world" and "outside world":
Mind is related to the *inside world* and matter to the *outside world*.
Then we can meet the following relations:
1. That level of reality which we register by *unconscious (non-intellectual)* objectivation is felt as independent from ourselves because it appears in front of us *without* our conscious action. This reality is felt to be independent from ourselves and we denote it with the "outside world".
2. That level of reality which we register by *conscious (intellectual)* objectivation is *not* felt as independent from ourselves because it appears *with* our conscious action. Therefore, this reality is *not* felt to be detached from ourselves and we denote it as the "inside world".

According to our conceptions, conscious and unconcious objectivation are analogous processes; various levels of the objectively existing (i.e., independent of the observer) *basic reality* are investigated on the basis of pictures.

An outside world is feigned

Therefore, the argument that the "hard" object belong to the *outside world* and the products of fantasy and scientific laws to the *inside world* is - as already mentioned above - obviously not correct; also the "hard" objects can only be registered by pictures and these pictures are located *in the head* as all pictures of the other levels of reality.

The brain obviously feigns an outside world for economic reasons; in this way the interplay between action and reaction is possibly more effective. The actual reality is identical with the *basic reality*, and we cannot make any statement about it.

The picture contains aspects of reality only in the form of symbols, i.e., the elements in the pictures are not identical with the corresponding elements in reality. In particular, the elements in the picture will not appear in reality at all.

The "world" which is designed by our brain is therefore an invention of our brain and we have seen in connection with the chick-experiment (Subsec. 1.2.5) that this invention is obviously strongly dependent on the species; the turkey has an optical conception of the chick which is obviously quite different from that of man.

1.4.7 Completions

Since thinking was developed relatively late by evolution, objectivation by thinking is therefore be less riped as the phylogenetically early developed unconscious objectivation by the constancy phenomena on the basis of the physiological apparatus. This explains the many uncertain and ambiguous results of thinking and in many cases no choice can be met between them. Therefore, doubts arise. Thus, the registration of reality by thinking still needs *reliability* (reproducible occurrence), and this veiled its concrete existence.

This lack of constancy is proved by a lot of false assessments which are regularly carried out by thinking individuals. (In the case of *unconscious* objectivation it is just this constancy which led us believe in the the real existence of "hard" objects.) That there is the possibility for such false assessments can be explained by the fact that in this phylogenetical phase the naked struggle for survival is no longer dominant; the individual is largely relieved from this pressure by the possibility of thinking. This produces individual freedom to act which can have -

according to the individual's decision - favourable as well as unfa-
vourable effects.

1.4.8 Consequences

We have stated in the last few sections that there is a strong analogy
between "hard" objects and products of mind (e.g., scientific laws). In
particular, we have outlined that all these "things" are registered by
objectivation and this means that they must *objectively* exist.

In the case of the "hard" objects, we have no difficulties in believing
that. However, in the case of the products of mind (e.g., scientific
laws) it is at first hard to conceive that because we are being accustomed to believe that the impressions in front of us are identical with
(basic) reality and not only with pictures of it. These pictures are ordered hierarchically on levels which we have called *levels of reality*.

1.4.9 Where Do the "Additions" Come From?

We have dicussed in the text above (see Fig.11 and the text relating to
it) that we can move from one level to another by means of "additions". However, their origins remain an open question. Due to the
many similarities between "hard" objects and products of mind, this
suggests the assumption that the products of mind (e.g., scientic
laws) are not always produced by the cognitive apparatus of the individual but they are "things" of reality, as it is the case of "hard"
objects. In other words, the similarity between "hard" objects and products of mind suggests that the products of mind (e.g., scientic laws)
are also *observed* by objectivation of *outside* information by the individuals, and the result is a *picture of reality* on a certain level.

The general law cannot be deduced from the special law!

When we base our considerations on the *principle of level-analysis*, it is no longer conclusive to deduce the laws which belong to a certain level (say level *A*) from another level (say level *B*) which is located in the hierarchy *below* level *A*. The structures (laws) of level *A* are by definition more *general* than those of level *B* and this means that level *B cannot* contain the "additions" which are necessary to pass from level *B* to level *A*. Otherwise the laws (structures) of level *A* are already contained on level *B*, i.e., level *A* cannot be located *above* level *B* in the hierarchy.

This is in accordance with Popper's statement [22] which we have already quoted in Subsec. 1.4.4 :

"To grope in all directions. Not that picture of science to gather and gather observations and from that the laws are distilled, so, as Bacon said, as wine is pressed from grapes. Within this conception the grapes are the observations and these grapes will be pressed and we obtain wine, i.e., the generalisation, the theory. This conception is utterly wrong ... "

In conclusion, according to our conception the laws cannot be deduced by *induction* but only by *deduction*, i.e., the *principle of level-analysis* only allows us to conclude from the structures of a certain level *A* to those of another level *B*, if level *B* is located *below* level *A*.

This means in particular that the creation of a new scientific law requires a new idea just at the beginning. From this idea follows the deduction and the comparison with well-known structures of levels which are below that level on which the new scientific law is established.

Example:

Newton's theory contains a completely new idea compared with Kepler's laws, namely the idea of *force*. It is impossible to deduce (neither deductive nor inductive) this new idea from Kepler's laws. From our point of view, Newton's concept of force can therefore be considered as an "addition", i.e., it is an "addition" to Kepler's laws.

The level of Newton's theory is therefore in the hierarchy above the level of Kepler's law; Kepler's laws can be found by deduction from Newton's theory. Newton himself believed mistakenly that he had only extended Kepler's laws by induction.

Where do the ideas come from?

Thus, the following questions arise: What is an idea? Where does it come from? Because of the strong analogy between "hard" objects and products of mind (e.g., scientific laws), which we have found above, we can establish ideas on certain levels of reality; they in general should not be components of the individual cognition apparatus but - as already outlined above - the cognition apparatus takes ideas from reality by objectivation. Then, the ideas are pictures of reality on certain levels of reality, just as "hard" objects are pictures of reality on a certain level.

Is the cognition apparatus an open system?

Such a process requires that the cognition apparatus is an *open* system, i.e., we have to permit that there is an interaction between the cognition apparatus and the environment. This looks strange at first, but this should not be the case because new developments indicate that there cannot be closed systems in physics at all. This is true within the framework of classical mechanics and is much more pronounced in quantum theory: All the quantum systems in the universe are interconnected to a giant, indivisible unit. Thus, why should the cognition apparatus be a closed system?

Inventions and discoveries

Therefore, the "additions" should in general be located in the outside world and not inside the cognition apparatus. This holds also in those cases when they are already located in the cognition apparatus when using them (effect of memory). From this point of view, we should consider "additions" as *discoveries*. In principle, we also have to permit that "additions" are processed by the cognition apparatus like a tree trunk which can be processed into a table; in this way we can change the world. Such processed "additions" can be considered as *inventions*.

1.5 Examples of the Principle of Level-Analysis

1.5.1 Levels of Reality within Classical Physics

The single levels within the framework of the principle of level-analysis are ordered by the degree of generality (Subsec. 1.4), where the level with a larger degree of generality is located above those with a lower degree of generality.

Hierarchy within classical mechanics

What can we say about the hierarchy in classical mechanics? It is natural at first to consider the phenomena on four levels:

1. *The level of assumptionless observations*
2. *The level of measuring instruments*
3. *Kepler´s level*
4. *The level of Newton´s physics*

We now want to examine each of these levels and construct a hierarchy with them.

1. Level of assumptionless observations

In assumptionless observations the things and phenomenon are ordered corresponding to the direct, visual impression, i.e., as they appear in front of us. Other information have by definition no place on this level.

All conclusions which are not in contradiction to assumptionless observations cannot be considered as wrong, even when that should be the case from a point of view of a higher level of reality.

For example, on this level it would not be consistent to state that the *geocentric world views* are wrong, and this is because the sun, moon and stars actually revolve around the Earth within the framework of assumptionless observations, i.e., the Earth seems to be in the centre on this level; such a simple view can be more useful (and therefore more favourable towards life) for certain purposes than a more sophisticated conception.

2. The level of measuring instruments

The observation of phenomena with measuring instruments means that more *precise and subtle* conclusions are in general possible than in the case of assumptionless observations. The information which we register by means of measuring instruments are in general more extensive and we can develop regions with them which are not accessible with the naked eye.

For example, in astronomy we can perform observations by means of a telescope, which are more *precise and subtle* than those with the naked eye in the case of assumptionless observations.

Therefore, statements about phenomena which can be made on the basis of a telescope are more general than those which are made within the frame of assumptionless observations. The level of assumptionless observations is therefore below the level of measuring instruments.

The astronomical data of the Dane *Tycho Brahe* (1546-1601) made a mark in history. The king of Denmark Friedrich II made it possible to set up an observatory on the island of Ven. Later Brahe became an astronomer with emperor Rudolf II of Prague. It is his merit to have increased the precision of measurements by the improvement of observation procedures. His experimental data on the precise positions of planets enabled *Johannes Kepler* (1571-1630) to deduce his laws.

3. Kepler's level

Emporer Rudolf II appointed not only Tycho Brahe but also Johannes Kepler, and Kepler worked in Prague as a mathematician. Unfortunately, the cooperation between Brahe and Kepler lasted only one year, i.e., from 19 October 1600 till Brahe's death (24 October 1601). However, it was sufficient to make Brahe's data available to Kepler. Brahe himself - in contrast to Kepler - was not capable in forming new ideas from his data. But Kepler was able to do that; he developed on the basis of Brahe's data his three famous laws which made modern astronomy possible.

Let us briefly summarize the three laws by Kepler:
1. The planet describes an ellipse with the sun at one focus.
2. The vector radius from the sun to the planet sweeps over
    equal areas in equal times.
3. The squares of the periodic times are proportional to the cubes of
    the major axes.

These three laws distinctly reflect that the geocentric world view no longer had a basis for Kepler. Here he was in contrast to Brahe who was still a proponent of a modified geocentric world view developed by himself.

Claudius Ptolemäus lived from 100 to 160, *Nikolaus Kopernikus* from 1473 to 1543. As is well known, both scientists developed systems of the world which played an essential role in the understanding of astronomical phenomena. The *heliocentric* system by Kopernikus was in many respects superior to the *geocentric* system by Ptolemäus. However, the heliocentric system was not so simple compared to the geo-centric system as is often stated. This is because both theories have to use complicated constructions (epicycles) which were sometimes awkward and confusing. In contrast to that *Kepler's ellipses* were admirably simple solutions.

Kepler concluded deductively

Kepler acted deductively, i.e., he concluded from the general to the special. From our point of view, the mathematical structures discovered by Kepler are ideas which are the result of a *conscious objectivation* (see Subsec. 1.4).

In other words, Kepler did not extract his three laws alone from Brahe's experimental data, as wine is pressed from grapes (here the grapes are Brahe's observations and the wine is the generalisation, i.e., Kepler's laws). In other words, Kepler did not find his laws by *induction*, i.e., the extraction of general features from special facts but - as already outlined above - by *deduction*.

In conclusion, in the case of Kepler's discovery (and probably in all the other cases which are similar to it), the *working hypothesis* is at the beginning and from that follows the deduction and the comparison with Brahe's data (see also [22]). The idea itself is given by the mathematical structures on which the three Kepler's laws are based or are reflected in these laws.

That level of reality, which is defined by the three Kepler's laws, has a larger degree of generality than that which is defined by Brahe's data; Kepler's level is therefore above that level on which measured observations are established and of course above the level of assumptionless observations.

Kepler's laws are valid for *all* orbits, i.e. also for those motions which are not or not yet realized. Kepler's ideas go beyond Brahe's *single* observations (which refer only to the concretely observed case) and is therefore more *general*.

The "additions" (see Fig. 11 and the corresponding text), which define the difference between Brahe's level and Kepler's level, are given by the mathematical structures of Kepler's three laws.

The ellipse as a solution was a new idea

One point should be briefly discussed, namely the form of the trajectories. What considerations led Kepler to the *ellipse* as a solution? It was a fundamental new idea which could not be deduced from the *circle* by *induction*. The ellipse was as an idea preconceived and was proved on the basis of experimental data. Kepler was convinced of the harmony of the world. Perhaps it was this conviction which helped him to *objectify* the ellipse as a trajectory for planets. The following comment by Popper is instructive [22]:

"Kepler used different forms (a pear-shaped form and various other forms) and finally he got a form which is very similar to the circle, namely the ellipse. That led to Kepler's first law. This was a very important step in many respects. The motion on a circle was easy to understand - Galilei only believed in the motion on a circle. Why the motion on a circle is so easy to understand can be recognized in the case of a wheel. One turns a wheel and the wheel continues on this motion; just as the law of inertia there is also an inertia of rotation; a wheel which is in frictionless motion keeps on turning (conservation of angular momentum). This is the intuitive basis of the theory for planet-motions on a circle, also that of Galilei's theory, and Galilei never gave up this theory. But Kepler remarked: < What a fool I was! > and he accepted the refutation."

("Kepler hat verschiedene Figuren, birnenförmige Figuren, verschiedene an-

dere Figuren verwendet und ist schließlich zu einer Figur gekommen, die dem Kreis ähnlich ist, nämlich die Ellipse. Das führt zum ersten Keplerschen Gesetz. Das war ein ungeheuer wichtiger Schritt in sehr vieler Hinsicht. Die kreisförmige Bewegung war leicht verständlich - Galilei hat nur an die kreisförmige Bewegung geglaubt. Warum die kreisförmige Bewegung so leicht verständlich ist, sieht man an einem Rad. Man stößt ein Rad an, das Rad dreht sich weiter, und ebenso wie das Trägheitsgesetz, gibt es hier ein Trägheitsgesetz der Rotation, die Erhaltung des Drehmoments, wenn man sich ein Rad vorstellt, das reibungsfrei läuft, so läuft es weiter und weiter. Das ist die intuitive Basis der Theorie kreisförmiger Planetenbahnen, auch der Theorie Galileis. Galilei gab diese Theorie nie auf. Aber Kepler schrieb: "Was für ein Dummkopf ich doch war!" und akzeptierte die Widerlegung.")

4. The level of Newton´s physics

Without any doubt, *Newton´s mechanics* and its concepts have been of determining relevance for the entire physics.

Newton´s mechanics contains a completely new idea, namely the *force,* which is not an element within Kepler´s theory. Newton lived from 1643 to 1727, i.e., he was born thirteen years after Kepler´s death.

It was and is not possible to deduce inductively or deductively this idea from Kepler´s theory. Newton´s force can be considered as an "addition" to Kepler´s laws, resulting in a completely new world view. Newton´s theory does not *supplement* Kepler´s theory but it is completely *superseded* by Newton´s world view.

Newton concluded deductively also

Kepler´s laws can be deduced from Newton´s theory; Newton´s mechanics also shows that Kepler´s theory is only approximately valid; it is valid as long as one can neglect the planetary mass in comparison to the mass of the sun. Furthermore, those effects are neglected within Kepler's theory, which are due to the attractive forces between the planets. One also knows that there not only ellipses are possible but also other trajectories (hyperbolas, parabolas) which follow from Newton's mechanics but not from Kepler's theory.

From all this follows that Newton's mechanics is more general than Kepler's theory. Therefore, the level of Newton's mechanics is above Kepler's level (see Fig. 12).

At first Newton believed mistakenly that he had only extended Kepler's theory by induction. However, Kepler's laws can be deduced from Newton's mechanics by deduction.

Because of its large success Newton's mechanics has been accepted as scientific revelation, and this is true without essential limitations up till today.

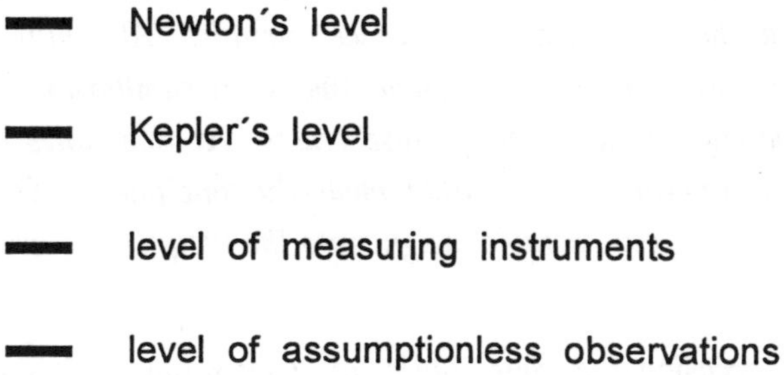

━━ Newton's level

━ Kepler's level

━ level of measuring instruments

━ level of assumptionless observations

Figure 12. Example for a hierarchy of levels of reality. The levels are vertically arranged according to the degree of generality.

1.5.2 Arthur Eddington's View

Concerning this topic, a appropriate example was given by the physicist *Arthur Eddington* (1882-1944). The example and the comments given are so clearly formulated that we want to quote the passage verbatim [24]:

"If the external forces p_{yy}, p_{xy} be given multiples of e^{ikx+at}, where k and a are prescribed, the equations in question determine A and C, and thence, by (9) the value of h . Thus we find

$$\frac{p_{yy}}{g\rho\eta} = \frac{(\alpha^2 + 2vk^2\alpha + \sigma^2)A - i(\sigma^2 + 2vkm\alpha)C}{gk(A - iC)} \ ,$$

$$\frac{p_{xy}}{g\rho\eta} = \frac{\alpha}{gk}\frac{2ivk^2A + (\alpha + 2vk^2)C}{(A - iC)} \ ,$$

where s^2 has been written for $gk+T'k^3$ as before ...

And so on for two pages. At the end it is made clear that a wind of less than half a mile an hour will leave the surface unruffled. At a mile an hour, the surface is covered with minute corrugations due to capillary waves which decay immediately, the disturbing cause ceases. At two miles an hour, the gravity waves appear. As the author modestly concludes, < Our theoretical investigations give considerable insight into the incipient stages of wave-formation >.

On another occasion the same subject of "Generation of Waves by Wind" was in my mind; but this time another book was more appropriate, and I read:

There are waters blown by changing winds to laughter

And lit by the rich skies, all day. And after,

Frost, with a gesture, stays the waves that dance

And wandering loveliness. He leaves a white

Unbroken glory, a gathered radiance,

A width, a shining peace, under the night.

The magic words bring back the scene. Again we feel Nature drawing close to us, uniting with us, till we are filled with the gladness of the waves dancing in the sunshine, with the awe of the moonlight on the frozen lake. These were not moments when we fell below ourselves. We do not look back on them and say, < It was disgraceful for a man with six sober senses and a scientific understanding to let himself be deluded in that way. I will take Lamb's Hydrodynamics with me next time >. It is good that there should be such moments for us. Life would be stunted and narrow if we could feel no significance in the world around us beyond that which can be weighed and measured with the tools of the physicist or described by the metrical symbols of the mathematician.

Of course it was an illusion. We can easily expose the rather clumsy trick that was played on us. Aethereal vibrations of various wave-lengths, reflected at different angles from the disturbed interface between air and water, reached our eyes, and by photoelectric action caused appropriate stimuli to travel along the optic nerves to a brain-centre. Here the mind set to work to weave an impression out of the stimuli. The incoming material was somewhat meagre; but the mind is a great storehouse of associations that could be used to clothe the skeleton. Having woven an impression the mind surveyed all that it had made and decided that it was very good. The critical faculty was lulled. We ceased to analyse and were conscious only of the impression as a whole. The warmth of the air, the scent of the grass, the gentle stir of the breeze, combined with the visual scene in one transcendent impression, around us and within us. Associations emerging from their storehouse grew bolder. Perhaps we recalled the phrase "rippling laughter".Waves --- ripples --- laughter --- gladness --- the ideas jostled one another. Quite illogically we were glad; though what there can possibly be to be glad about in a set of ae-thereal vibrations no sensible person can explain. A mood of quiet joy suf-fused the whole impression. The gladness in ourselves was in Nature, in the waves, everywhere. That's how it was.

It was an illusion. Then why toy with it longer? These airy fancies which the mind, when we do not keep it severely in order, projects into the external world should be of no concern to the earnest seeker after truth. Get back to the solid substance of things, to the material of the water moving under the pressure of the wind and the force of gravitation in obedience to the laws of hydrodynamics. But the solid substance of things is another illusion. It too is a fancy projected by the mind into the external world. We have chased the solid substance from the continuous liquid to the atom, from the atom to the electron, and there we have lost it. But at least, it will be said, we have reached something real at the end of the chase - the protons and electrons. Or if the new quantum theory condemns these images as too concrete and leaves us with no coherent images at all; at least we have symbolic co-ordi-nates and momenta and Hamiltonian functions devoting themselves with single-minded purpose to ensuring that $qp-pq$ shall be equal to $ih/2\pi$."

"We have torn away the mental fancies to get at the reality beneath, only to find that the reality of that which is beneath is bound up with its potentiality of awakening these fancies. It is because the mind, the weaver of illusion, is also the only guarantor of reality that reality is always to be sought at the base of illusion. Illusion is to reality as the smoke to the fire. I will not urge that hoary untruth < There is no smoke without fire >. But it is reasonable to inquire whether in the mystical illusions of man there is not a reflection of an underlying reality."

Comments to Arthur Eddington's view

It is relatively easy to bring Arthur Eddington's view in line with our conception, namely our result that we never can observe *basic reality* but only pictures of it on various *levels of reality*.

First the production of waves by wind is mentioned; waves and wind are elements which are established at the "level of everyday life" if we restrict ourselves to the picture which is designed *unconsciously* by the physiological apparatus. On the basis of this situation Arthur Eddington brings two facets of one and the same thing (wave and wind) into play:

1. *Hydrodynamics* which allows a pure scientific consideration.

2. A *product of literary fantasy* which seemed appropriate to Eddington in a certain situation of life. In other words, the product of fantasy was felt to be *favourable towards life* in that certain situation. (As we have discussed frequently above, in nature not cognition plays the essential role but the differentiation between "favourable towards survival" and "hostile towards survival". Therefore, it is important to look for situations which are favourable for life or, on the other hand, to produce such situations.)

Then, Eddington dicusses that *both* conceptions (hydrodynamics, product of fantasy) are illusions. What does that mean? This means from our point of view that neither of the two pictures contain elements of the basic reality. Eddington also considers the hydrodynamical picture as illusion, because the matter was chased out of the uniform liquid into the atom, and out of the atom into the electron, then it got lost.

In other words, three different scientific illusions arise in the sequence

uniform liquid - atom - electron.

Eddington considers these pictures as illusions and this is because three pictures exist alongside each other, and Eddington rejects none of them, i.e., none of them are considered to be wrong.

This can only be reduced to a common denominator when we apply the *principle of level-analysis* which we have developed above:
None of the so-called illusions represents the *basic reality*; they are *pictures* of reality on different levels (see Fig. 10) which are ordered according to the degree of generality.

In summary, we may conclude that our conception is fully reflected in Eddington´s view.

1.6 Diversity of Pictures

1.6.1 The Space of Everyday Life

Space-time only arises in the real context

The picture and also its "frame", space and time, is located in the head of the observer. We know from experience that space-time arises only in connection with objects and processes, i.e., an *empty* space-time is not observable.

That obviously means that space-time is not installed in the brain as a definite entity but is only "inserted" there if there is actually something to represent, i.e., when our sense organs register objects and processes from the reality outside. (Here we at first talk only about the space of everyday life; generalisations will be discussed below.)

Dependency on the species

Because of these features we have to assume that space-time is dependent on the biological structure of the individual and, therefore, it will in general be dependent on the *species*.

From the dramatic and deeply shocking results of the chick experiment (Subsec. 1.2.5) we have concluded that the turkey must optically experience the world quite differently from us, although the eyes of the turkey are quite similar to ours. There is obviously no similarity between what the turkey experiences and the conception the human observer has in his mind in the same situation!

But when the pictures of both species are so different from each other, we cannot exclude that their frames of representation are also different from each other.

How the frame of representation is designed in the case of the turkey cannot be stated at present. But in general we have to assume that the frame of representation is also involved in the evolutionary process. The frame of representation of the turkey has by no means to be an Euclidean space, and it must not have three dimensions at all. Evolution will have developed such a frame of representation that is *useful* and *favourable towards survival* for the turkey. That is the only criterion, at least in the early phase of evolution, and everything following that is based on these early developed features. The situation is summarized once more in Fig. 13.

1.6.2 Other Spaces

Let us come back to the space of everyday life of man. In this case we also have to assume that space has been developed by evolution; whether or not it is still involved in an evolutionary process remains an open question at present. But evolution has created the possibility for *thinking*, and this evolutionary step is obviously accompanied with a further development of the space of everyday life; this is due to the following reason:

The space of everyday life is an element of unconscious objectivation, i.e., it will be constructed by the brain without our conscious action (see also Subsec. 1.4.5). However, the "objects" of another level of reality, which are registered by conscious objectivation, i.e., with our conscious action, are in general represented within a frame of representation which is different from the space of everyday life.

In *mathematics*, a diversity of spaces has actually been developed. Also in physics the space of everyday life is not sufficient for the decription of phenomena; the spaces of the theory of relativity (the STR, GTR; see also Subsec. 1.1) deviate considerably from the space-time of everyday life. The space-time of everyday life, which is Euclidean in a good ap-

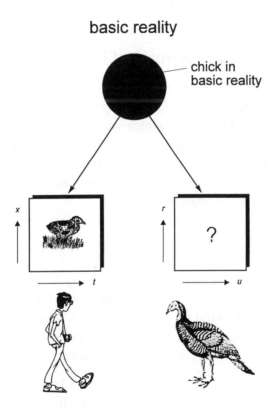

Figure 13. The frame of representation is possibly dependent on the species. In the case of man, the frame of representation is on the "level of assumptionless observations" given by the space of everyday life (x, t). The frame of representation for the turkey (marked by the coordinates r and u) can however be completely different from ours. We know nothing about the picture of the chick in the head of the turkey. Furthermore, both man and turkey can say nothing about the chick in basic reality.

proximation, is obviously no longer sufficient when we analyse phenomena on higher levels of reality.

At this point we do not want to continue this discussion but will take it up again in the following sections.

1.6.3 Alternatives to the Space of Everyday Life

We have stated above that the space-time of everyday life comes into play without our conscious action; this space-time is used for the ordering of objects and processes which we have directly in front of us and which are experienced without auxiliary elements as measuring instruments. Are there *alternatives* to the space-time of everyday life? In other words, does a frame of representation exist which is equivalent to the space-time of everyday life?

Equivalent informations

Man is certainly tied in his behaviour to the space-time of everyday life, but other biological systems could have developed in evolutionary processes another kind of frame of representation which could be equivalent (or approximately equivalent) to the space-time of everyday life with respect to the contents of information.

Do such alternatives exist? Although we do not know, we have to assume that nature does not preclude such a possibility. On higher levels of reality, there definitely exist such possibilities. Why not on the level of everyday life as well?

As is well known, there exist *mathematical* transformations which allow one to transform the contents of the space-time of everyday life into another frame of representation without loss of information (see also the scheme in Fig. 14).

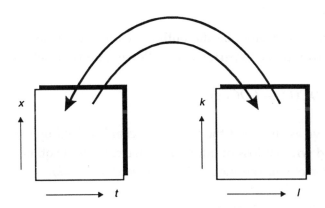

Figure 14. Two frames of representation which are equivalent concerning their information (for simplicity, we want to consider a space with only one dimension). Does there exist a mathemetical tranformation which tranforms the information in one space (characterized by coordinate x and time t) into another space (characterized by k and l) and vice versa?

1.6.4 Fourier-Space

The Fourier transform allows such an operation. In Fig. 15a, the outline of a locomotive is shown which is represented in the space of everyday life within two dimensions. (For simplicity, the problem is only treated in two dimensions. Furthermore, we do not want to analyse sequences of motion and that also means time t need not appear.) The outline of the locomotive is mathematically described as a function of

x which we want to call $LOC(x)$.

The transition from one frame to another means that, instead of x, we have another coordinate, say k. In this new space, the outline of the locomotive can be described as a function of k which we want to call $LOC(k)$.

Our goal is to describe the outline $LOC(k)$ by means of $LOC(x)$, that is, we would like to perform the following transformation:

$$LOC(k) \leftarrow LOC(x) \ . \tag{8}$$

Since we assume that the transition, characterized by (8), is to be performed without loss of information we can, on the other hand, describe inversely the outline $LOC(x)$ by means of $LOC(k)$:

$$LOC(x) \leftarrow LOC(k) \ . \tag{9}$$

So much as to the formal connection, as it is also given in Fig. 14.

Within the frame of the *Fourier transform* the transition (8) is given by

$$LOC(k) = \int_{-\infty}^{\infty} LOC(x)\exp(-ikx)dx \ . \tag{10}$$

On the other hand, the transition (9) is given by the following relation:

$$LOC(x) = \int_{-\infty}^{\infty} LOC(k)\exp(ikx)\frac{dk}{2\pi} \ . \tag{11}$$

When we now put in the numerical values for $LOC(x)$ $[LOC(x)=LOC(-x)]$ into Eq. (10) we obtain Fig. 15b. On the other

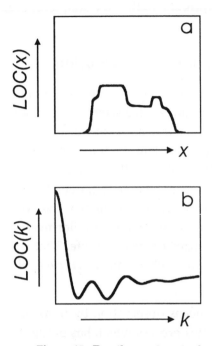

Figure 15. Details are given in the text.

hand, when we know the numrerical values for the outline $LOC(k)$, we can calculate $LOC(x)$ by means of Eq. (11).

There is no similarity between the locomotive in Fig. 15a and that in Fig. 15b, although the contents of information in both pictures are identical. However, the frames of representation are distinctly different from each other. The locomotive in Fig. 15a is represented within the space of everyday life; here, coordinate x has the dimension of a length (for, example, metre). In the case of Fig. 15b, the locomotive is represented within the so-called Fourier-space. The Fourier-space is also called reciprocal space and this is because coordinate k has the dimension of a reciprocal length (for example, 1/metre).

1.6.5 Spaces are Tailored by Evolution

Although the form in Fig. 15b is quite different from that in Fig. 15a, it also represents the locomotive. This seems to be absurd at first, but it is not so because we are merely accustomed to the space of everyday life. But is it sufficient to explain the initial strangeness in the figure by an effect of habit? Probably not. It is probable that man´s biological structure excludes the Fourier-space as a frame of representation at the level of everyday life; we have to assume that the evolution has developed, together with the sense-organs, a tailor-made frame of representation which is the space of everyday life. That is the situation on the level of everyday life, where an *unconscious objectivation* takes place.

Conscious objectivation means observation by thinking (*principle of level-analysis*). It was just this process which has brought out the Fourier-space. In this respect, Fourier-space can also be thought as existing, even when it is not compatible with the world of the five senses.

Is the Fourier-space spontaneously realizable?

Is Fourier-space spontaneously realizable in the case of other species, such as the space of everyday life is spontaneously realized in the case of man? (Clearly, here "spontaneously realized" again means that space comes into play without the conscious action of the observer.) We do not know that. Therefore, we have to restrict ourselves to some general remarks and tendencies.

It is difficult for us to state something about the cognition apparatus of other biological systems, for example, as in the case of the turkey.

Although we already learned a lot through the chick experiment, we know nothing about the structure of the frame of representation of the turkey and how it orders the relevant objects and processes. The only definite statement we can make at the moment is that the turkey must

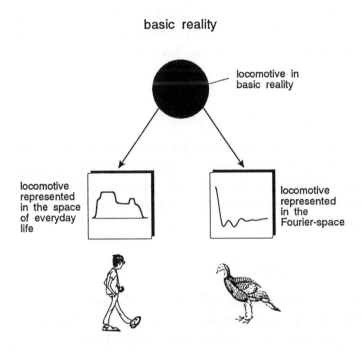

Figure 16. The locomotive is embedded within *basic reality*, so as everything is embedded in it. The objects and processes of *basic reality* can in principle not be experienced directly by biological systems (e.g., man and turkey). But we can form pictures of them, pictures of reality at different levels.

The picture of the locomotive in the space of everyday life has a normal structure for us, but only for us. If we speculate that the turkey orders the objects and processes within the Fourier-space, the turkey would form a picture from the same locomotive which is quite different from ours. The situation could correspond to that in the chick experiment when we replace the locomotive by the chick (see also Fig. 13).

have a picture in front of it which is quite different from ours. On the basis of various arguments, we have recognized above that the frame of representation does depend on the species, so we have good reason to assume that not only the picture but also the frame of representation for the picture should be quite different from ours, because the frame of representation is also probably tailor-made by evolution. Why should everything else be shaped by evolutionary processes but not the frame of representation? Such an assumption would be most unlikely.

Pictures in Fourier-space

Although we do not know how the cognition apparatus of the turkey is structured, for our further discussion it will be very illuminating to speculate that within the cognition apparatus of the turkey the Fourier-space is realized and not our space of everyday life. The situation is discussed in more detail in Fig. 16.

Changes in the pictures

Let us first investigate how changes, which can be made in the space of everyday life, are reflected in Fourier-space. In Fig. 17a, we recognize immediately that the locomotive (Fig. 17b) has no funnel. If we now transform this locomotive without a funnel into the space of the turkey (Fourier-space) we recognize that the picture has hardly changed compared to Fig. 17d (locomotive with a funnel) and this is because the local change in the space of everyday life is smeared in Fourier-space over the whole picture, i.e., *local* effects in space of everyday life become *non-local* in Fourier-space.

Why the turkey does not recognize its chick

In conclusion, local changes in the space of everyday life can be less pronounced in Fourier-space. Due to such an effect, it might be possible that within the turkey's space - which we have taken, on a trial basis, as Fourier-space - it is hardly able to distinguish between the chick and the weasel. The experiments by Wolfgang Schleidt actually show that the turkey obviously can only identify its chick if *specific acoustical* signals are available simultaneously. This fact becomes immediately understandable when we assume that the frame of representation of the turkey is given by Fourier-space or a space which produces similar effects. However, it can of course be also quite different; we still know very little about other species.

Different positions, sequences of motion

In Fig. 18, the locomotive is shown in three different positions. All three situations have been transformed by Fourier transform into the space of the turkey. Now we observe distinct changes which are, however, again *non-local* in character and this is of fundamental importance for the concept of *motion in Fourier-space* and, furthermore, for the concept of *velocity*.

In the space of everyday life, motion is definable as local changes. In the case of an uniform motion the velocity v of an object (for example, the locomotive) is - as is well known - given by the position x_1 at time t_1 and the position x_2 at time t_2 as follows:

$$v = \frac{x_2 - x_1}{t_2 - t_1} \ . \tag{12}$$

Such a kind of velocity is not definable in Fourier-space since we

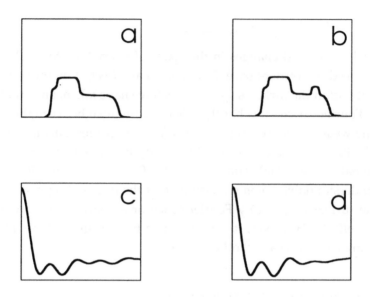

Figure 17. A locomotive with and without a funnel; a and b: Representation in the space of everyday life; c and d: Representation in Fourier-space.

cannot recognize in Fig. 18 the usual kind of motion. In contrast to the space of everyday life, where the picture is changed only locally (the locomotive drives from one position to another), in the picture of Fourier-space *non-local* changes occur, i.e., *all* positions in Fourier-space are affected by the *local* changes in the space of everyday life, and this makes a definition of velocity in accordance with (12) impossible.

However, such a kind of velocity belongs to the fundamental elements of many theories as, for example, in Newton´s mechanics and the theory of relativity. For all these theories the conceptual frame would be abolished in Fourier-space. Although the contents of information is exactly the same in both spaces, the world in Fourier-space is not comparable with that in the space of everyday life.

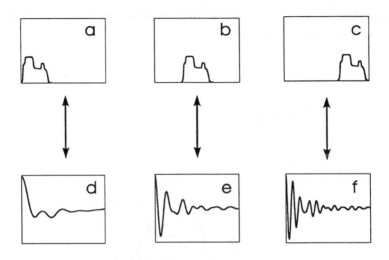

Figure 18. The locomotive in different positions in the space of everyday life (a, b, c) as well as the corresponding pictures in Fourier-space (d, e, f).

No possibility for relating

Certain positions in the space of everyday life are not related to certain positions in Fourier-space (the space of the turkey). There is no possibility for relating; the funnel in Fig. 17b cannot be recognized in Fig. 17d although that picture completely contains the funnel; it is smeared over the whole picture.

The reason for this property is that each *single* point of the space of everyday life contains the information of *all* points of Fourier-space (see also Fig. 19a) and vice versa: Each *single* point of Fourier-space contains the information of *all* points of the space of every day life (see also Fig. 19b). This is the reason why the pictures in Fig. 15 cannot be compared to each other, although they have the same contents of information. Both pictures are correct but their contents of information is organized differently.

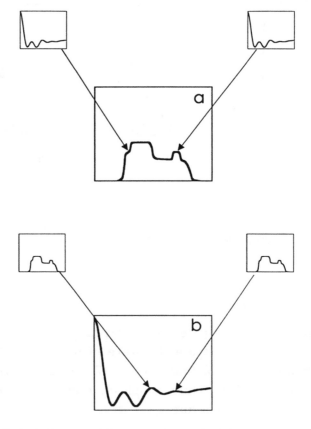

Figure 19. Each *single* point of the space of everyday life contains the information of *all* points of Fourier-space, which is demonstrated in (a) by two examples. On the other hand, each *single* point of Fourier-space contains the information of *all* points of the space of everyday life which is also demonstrated by two examples (in (b)).

Classification in accordance with the principle of level-analysis

We can conclude from our discussion, that we have at least two *equivalent* frames of reference (the space of everyday life, Fourier-space) which however are quite different from each other concerning their structure. In principle, the pictures could be projected on the space of everyday life as well as on Fourier-space (see also Fig. 20) which however would be very strange to us, since all our conceptions are tailor-made to the space of everyday life. The evolution has at an early stage fixed itself on a specific type of frame of representation, which it has further developed up to the present state. On the level of everyday life, where *assumptionless observations* take place (*unconscious objectivation*), one cannot switch over to another frame of representation as, for example, to the Fourier-space; such a process is not planned by nature. However, in the case of *thinking*, i.e., in the case of *conscious objectivation*, it is possible to switch over from one frame of representation to another. So we know, besides the space of everyday life and the Fourier-space, there is a diversity of other spaces which become important on other levels of reality; for example, high-dimensional spaces, curved spaces, complex spaces, etc. Each of these spaces is used in specific physical theories. Already this fact agrees completely with the foundations of the *principle of level-analysis*.

Remark concerning the situation in quantum theory

Within Fourier-space we obviously cannot define a *velocity* in the sense of Eq.(12). The reason for this fact is that there does not exist a *trajectory* for the locomotive in Fourier-space, such as it exists within the space of everyday life on which Eq. (12) is based.

In this connection, it is also interesting to note that within *quantum theory* no trajectory for particles can exist and therefore no velocity in the sense of Eq. (12). But this does not immediately mean that quantum phenomena are projected on Fourier-space. However, it is shown in [5] that Fourier-space obviously plays an important role in quantum theory.

That within quantum theory the concepts of velocity and trajectory are not definable is firmly established. In [25] the following is quoted:

"This circumstance shows that, in quantum mechanics, there is no such concept as the velocity of a particle in the classical sense of the word, i.e., the limit to which the difference of the coordinates at two instants, divided by the interval Δt between these instants, as Δt tends to be zero."

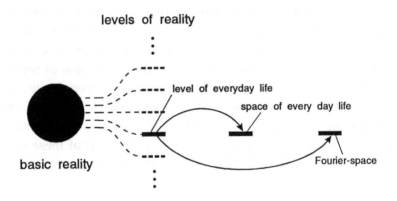

Figure 20. With respect to the contents of information, Fourier-space is completely equivalent to the space of everyday life. Therefore, the objects and processes on the level of everyday life could in principle also be projected on the Fourier-space. Both spaces have therefore in the hierarchy of levels of reality the same vertical position. However, on the level of everyday life, where *assumptionless observations* take place (*unconscious objectivation*), one cannot switch over to another frame of representation as, for example, to the Fourier-space; such a process is not realizable in nature. It remains of course an open question whether or not such spaces are realizable in connection with other species as, for example, in the case of the turkey.

This holds not only with respect to the velocity, but the *non-local* effect, which appears when one proceeds from the space of everyday life to Fourier-space, is obviously also a typical feature of quantum theory.

The funnel, which is well seperated from the other parts of the locomotive in the space of everyday life, is mixed in Fourier-space with the other parts of the locomotive, so that the locomotive can no longer be divided into the usual seperate elements, i.e., *non-local* effects appear in the transition from the space of everyday life to Fourier-space.

As mentioned, certain effects in quantum theory are obviously also non-local in character; they can obviously only be avoided by the assumption that the present form of quantum theory is not correct (see, for example, also [26]). In [5], it is discussed that this quantum weirdness is actually due to a Fourier-space effect, similar as in the case of the locomotive.

1.7 Other Spaces, Other Particles

Change of frame of representation at the atomic level

The locomotive is a macroscopic object. It appears as a *continuum* on the level of everyday life, and its shape is fixed by the boundary to space. Continuum means unbroken packed matter; this is the impression within the frame of assumptionless observation at the level of everyday life.

However, when we look more precisely by means of measuring instruments, i.e., with larger resolution, matter has an *atomic* structure: The closely- packed matter appears on this level as *empty* space with a slight addition of real substance, and this is of course also true for the locomotive at this level.

The construction of equivalent but fundamentally different pictures of the *same* object (see Fig. 15) is not only possible at the level of everyday life but also at the *atomic level*; also in this case we can put *equivalent* information into *different* frames of representation. In this way we can produce "worlds" at the *atomic level* which are qualitatively different from each other, although they are based on exactly the same contents of information. In the following, we would like to discuss this point in more detail.

1.7.1 The Harmonic Solid

The locomotive can be conceived as a large crystalline solid, and the properties of the locomotive at the atomic level are fixed by the properties of this crystalline solid. In Subsec. 1.5.2, we have discussed Arthur Eddington's view, and this view can also be applied here: Remember, the matter was chased out of the uniform liquid into the atom; so the matter is chased out of the uniform (unbroken packed) locomotive into the crystalline (atomic) solid. Let us examine the atomic solid within various frames of representation; for simplicity, let us do that by means of a very simple solid.

Hamiltonian in the harmonic approximation

Let us consider an ideal crystal which consists of a regular arrangement of N identical atoms of mass m. It is assumed that these atoms are fixed at equilibrium positions.

We want to decribe the displacements from the lattice positions by the Cartesian coordinates q_λ, $\lambda = 1,...,3N$. Then, the potential energy U depends on all these coordinates:

$$U = U(q_1,...,q_{3N}) \ .$$
(13)

The Hamiltonian (see, for example, [5]) for such a system is given by

$$H = \sum_{\lambda=1}^{3N} \frac{p_\lambda^2}{2m} + U(q_1,...,q_{3N}) \ ,$$
(14)

where the first term of the right-hand side is the kinetic energy. The quantities

$$p_\lambda = m\dot{q}_\lambda$$
(15)

are the momenta canonically conjugate to q_λ. We may expand the potential energy U in a Taylor series, if the displacements from the equilibrium positions are small in comparison to the atomic distances:

$$U = \frac{1}{2} \sum_{v,\lambda} \Phi_{v\lambda} q_v q_\lambda \ , \tag{16}$$

where the constant term in the series does not contribute and is therefore not quoted in (16). The linear term vanishes because it corresponds to the minimum of energy; the force on any atom vanishes in equilibrium. The first non-vanishing term is then given by (16). The coefficients $\Phi_{v\mu}$ are the second derivatives of U at the equilibrium positions. We would like to restrict ourselves to the harmonic solid, i.e., all higher-order terms in U are assumed to be negligibly small. Then, instead of (14) we obtain for the Hamiltonian in the harmonic approximation the following expression:

$$H = \sum_{\lambda=1}^{3N} \frac{p_\lambda^2}{2m} + \frac{1}{2} \sum_{v,\lambda} \Phi_{v\lambda} q_v q_\lambda \ . \tag{17}$$

Transformation to normal modes

Now we can perform the so-called orthogonal transformation. For this purpose we go from the space (q-space) with the particle coordinates

$$q_\lambda \ , \ \lambda = 1,...,3N$$

to another space (Q-space) which has the coordinates

$$Q_\rho \ , \ \rho = 1,...,3N \ .$$

The quantities q_λ and Q_ρ are related by the following expression:

$$q_\lambda = \sum_{\rho=1}^{3N} v_{\lambda\rho} Q_\rho \ . \tag{18}$$

Let us assume that the coefficients $v_{\lambda\rho}$ satisfy the condition of orthogonality:

$$\sum_\lambda v_{\lambda\rho} v_{\lambda\sigma} = \sum_\lambda v_{\rho\lambda} v_{\sigma\lambda} = \delta_{\rho\sigma} \ . \tag{19}$$

Equation (19) guarantees that the coordinates Q_ρ and q_ν are only rotated relatively to each other. The further condition

$$c_\rho v_{\nu\rho} = \sum_\lambda \Phi_{\nu\lambda} v_{\lambda\rho} \tag{20}$$

causes that this rotation is just into the position of the main axes of the potential energy (see, for example, also [27]); the transformations which are given by the Eqs. (18) - (20) are called "transformations to normal modes". Equation (20) connects the new coupling parameters c_ρ to the old one $\Phi_{\nu\lambda}$ and coefficients $v_{\lambda\rho}$. With Eqs. (18) - (20) we then obtain

$$\frac{1}{2}\sum_{\nu,\lambda} \Phi_{\nu\lambda} q_\nu q_\lambda = \frac{1}{2}\sum_{\nu,\lambda} \Phi_{\nu\lambda} \sum_\sigma v_{\nu\sigma} Q_\sigma \sum_\rho v_{\lambda\rho} Q_\rho$$

$$= \frac{1}{2}\sum_{\nu,\sigma,\rho} c_\rho v_{\nu\rho} v_{\nu\sigma} Q_\sigma Q_\rho = \frac{1}{2}\sum_{\sigma,\rho} c_\rho \delta_{\rho\sigma} Q_\sigma Q_\rho$$

$$= \frac{1}{2}\sum_\rho c_\rho Q_\rho^2 \ . \tag{21}$$

The form for the expression of the kinetic energy remains unchanged within the space with coordinates Q_ρ and we get instead of Eq. (14)

$$H = \sum_{\rho=1}^{3N} \left\{ \frac{P_\rho^2}{2m} + \frac{1}{2} c_\rho Q_\rho^2 \right\} \quad , \tag{22}$$

where

$$P_\rho = m\dot{Q}_\rho \quad , \quad \rho = 1,...,3N \tag{23}$$

are the momenta canonically conjugate to Q_ρ.

The Hamiltonian, which is given in the space with coordinates Q_ρ, appear as $3N$ terms which are completely independent from each other; each term has the form

$$H_\rho = \frac{P_\rho^2}{2m} + \frac{1}{2} c_\rho Q_\rho^2 \quad . \tag{24}$$

As is well known, H_ρ is the Hamiltonian for a harmonic oscillator.

1.7.2 Comparison of the Two Solid State Pictures

1. Results given in the space with coordinates q

Here we have N particles with $3N$ coordinates which are coupled to each other due to the potential energy U. The Hamiltonian within the frame of the harmonic approximation is given by [Eq.(14)]

$$H = \sum_{\lambda=1}^{3N} \frac{p_\lambda^2}{2m} + \frac{1}{2} \sum_{v,\lambda} \Phi_{v\lambda} q_v q_\lambda \quad .$$

One gets the trajectories in q-space by solution of Hamilton´s equations (see, for example, [5]):

$$\dot{q}_i = \frac{\partial H}{\partial p_i} \ , \ \dot{p}_i = -\frac{\partial H}{\partial q_i} \ ; \ i = 1, 2, \ldots, s \ .$$

Picture of the solid in q-space:
We have interacting particles, i.e., particles which are not independent of each other. The existence of the potential energy means that we cannot attribute a definite energy to the particles, since part of the energy is *between* the particles as potential energy so to speak. In other words, the individual particles, which build up the solid in q-space, are not well defined. Clearly, the self-energies of the particles are much larger than the energies due to the interactions, so that the particles are defined in a good approximation, but only in good approximation and not exactly.

2. Results given in the space with coordinates Q

Here we have $3N$ non-interacting oscillators, i.e., the oscillators are independent of each other.

The Hamiltonian [Eq. (22)]

$$H = \sum_{\rho=1}^{3N} \left\{ \frac{P_\rho^2}{2m} + \frac{1}{2} c_\rho Q_\rho^2 \right\}$$

is separated into $3N$ terms. The equations of motion for the oscillators remain unchanged, i.e., we obtain in Q-space relations analogous to that in q-space:

$$\dot{Q}_i = \frac{\partial H}{\partial P_i} \;,\; \dot{P}_i = -\frac{\partial H}{\partial Q_i} \;;\; i = 1,2,...,3N \;.$$

Picture of the solid in Q-space:
We have single non-interacting oscillators. The concept of "potential energy" does not appear and this means that we can talk in terms of defined units in this case.

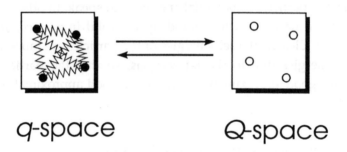

q-space   Q-space

Figure 21. The solid is, in q-space, a lattice which consists of particles (full circles); the particles interact, and this process is symbolized in the figure by springs.
On the other hand, the *same* solid consists in Q-space of oscillators (open circles) and there is no interaction between them. The concept of potential energy does not appear.
There is no direct relation between a *single* oscillator in Q-space and a *single* particle in q-space, but the *ensemble* of the $3N$ oscillators is completely equivalent to the *ensemble* of the $3N$ particles.

3. Discussion

The situation is summarized in Fig. 21. We have always to consider that a *single* oscillator cannot be transformed into a *single* particle and vice versa; there exists no direct relation between the single oscillator in Q-space and the single particle in q-space because *each* coordinate of a particle is connected to *all* coordinates of the oscillators [see Eq.(18)] and vice versa (non-local effects).

However, *all* oscillators are equivalent to *all* lattice particles concerning their contents of information. The transition from Q-space to q-space means that there are different representations but the spaces

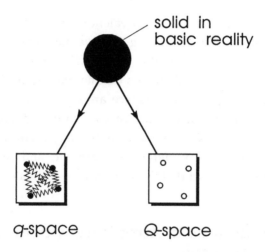

Figure 22. The solid is embedded in *basic reality*, so as is everything embedded in it. The objects and processes of *basic reality* cannot in principle be experienced directly. But we can form pictures of them, pictures of reality at different levels.

The picture of the solid in q-space, which is identical with the space of everyday life, has the usual structure for us and this is because our intuition is tailor-made for this space. However, were our intuition tailor-made for Q-space, the picture in front us would show the same solid in a quite different arrangement, although the contents of information of q-space is identical with that of Q-space.

The situation given in this picture corresponds completely to that in Fig. 16.

themselves are equivalent because the same contents of information can be put into both spaces, i.e., either into Q-space or into q-space.

This is quite similar to the case which we have discussed in connection with the locomotive: The transformation from the space of everyday life to the Fourier-space has led to quite different, but equivalent pictures. Also here, we have observed that elements, e.g., the funnel, which are localized in the space of everyday life are smeared in the Fourier-space, i.e., such elements are not localized in Fourier-space and we are confronted with *non-local* effects.

From the different nature of equivalent spaces, which definitely exist (see, for example, Fig. 16 and Fig. 22), we have to conclude that none of them can be realized within *basic reality* and this is because the *equivalence* of the spaces demands that no one can be favoured. So, on the one hand, q-space and Q-space are equivalent (Fig. 21) and, on the other hand, the space of everyday life and Fourier-space are equivalent (Fig. 15). None of these spaces may occur in *basic reality* because this would be preferential treatment, and this would be in contradiction to the fact that these spaces are equivalent.

1.7.3 Classification in Accordance with the Principle of Level-Analysis

When we arrange both conceptions (particles in q-space, oscillators in Q-space) in accordance with the principle of level-analysis, we obtain Fig. 22 which is completely equivalent to Fig. 16. We always have to keep in mind that neither the particles nor the oscillators are embedded "in" space; within the particular pictures "only" *geometrical positions* occur; they are projections on q-space and Q-space, respectively. The representations in the particular spaces have to be considered as "alternative pictures".

1.8 Levels of Reality and Mach's Principle

Space and time as absolute quantities

In Subsec. 1.1, we have studied fundamental aspects of space and time; it turns out that space and time in Newton's mechanics are *absolute* in character, which has to be considered as an essential shortcoming. Furthermore, we have outlined that this difficulty was not at all eliminated by the theory of relativity, i.e., the problem still exists up to the present day. In the following, we want to briefly summarize the essential arguments which show why *absolute* space-time is unphysical and therefore not acceptable.

Concerning the term "absolute" note the following:
1. Absolute space was invented by Newton for the explanation of *inertia*. However, we do not know of any other phenomenon for which absolute space would be responsible. So, the hypothesis of absolute space can only be proved by the phenomenon (inertia) for which it has been introduced. This is unsatisfactory and artifical.
2. The term "absolute" not only means that space is *physically real* but also *"independent in its physical properties, having a physical effect, but not itself influenced by physical conditions"* [6]. This must also be considered as unsatisfactory.

Both points indicate that absolute space is actually an *unphysical* quantity. Although Newton's mechanics was very successful (and it is still used in many calculations) a lot of physicists could not accept the concept of an absolute space. This is demonstrated by the fact that scientists tried to solve this problem again and again up to the present day.

Concerning absolute space, *Max Born* expressed his view very clearly. He wrote [7]:

"Indeed, the concept of absolute space is almost spiritualistic in character. If we ask, "What is the cause of centrifugal forces?" the answer is: "Absolute space." If, however, we ask what absolute space is and in what other way it expresses itself, no one can furnish an answer other than that absolute space is the cause of centrifugal forces but has no further properties. This consideration shows that space as the cause of physical occurrences must be eliminated from the world picture."

The elimination of space as an active cause

This is why Mach eliminated space as an active cause in the system of mechanics (Mach´s principle). To him, a particle does not move in unaccelerated motion relative to space (as in the case of Newton´s conception), but relative to the centre of all the other masses in the universe [6]; in this way, the series of causes of mechanical phenomena was closed, in contrast to Newton´s mechanics.

In fact, absolute space and, of course, absolute time must be considered as metaphysical elements because they are, in principle, *not accessible to empircal tests:*
There is no possibility of determining space coordinates $x=x_1$, $y=x_2$, $z=x_3$ and time t. We can only say something about distances in connection with masses, and times in connection with physical processes.

Even if one is generally not a follower of Mach´s philosophy one is not be able to get out to take *Mach´s principle* - formulated above - literally, because we have to accept that the space-coordinates and the time are not observable as independent quantities, i.e., the absolute character, which these quantities have within the framework of Newton´s mechanics and even in the theory of relativity, is in contradiction to experience!

Space and time as auxiliary elements

The usual conception of the world is that matter is
                    "embedded in" space and time.
In the Subsec. 1.2-1.7 we have outlined that this should not be the case:
A lot of facts indicate that reality is
                    "projected on" space and time.

On the basis of this new conception, we have deduced a new order-principle which we have called
                    principle of level-analysis.

Within this principle there is a
                    basic reality
which can in principle not be experienced. However, we can form
                    pictures
on various
                    levels of reality.
There will be no similarity between these *pictures of reality* and *basic reality*. The pictures and the levels of reality, respectively, are arranged according to the degree of generality.

Within such a scheme *Mach's principle* is of course satisfied; within the structures of a picture, no physical effects can take place: Within a picture there are never material objects but only symbols, and space and time cannot act on symbols; the symbols are ordered by means of space and time, i.e., space and time relate the symbols to each other.

The question whether or not space and time are also elements of *basic reality* must be answered in the negative. *Basic reality* contains the real something and, in principle, space and time could act on the real something if space and time were also elements of *basic reality*. However, as stated above, space and time cannot be elements of *basic reality* because we have to consider them as *picture-specific* quantities and, therefore, they cannot occur in *basic reality* by definition. Furthermore, we can say quite generally that there is no *picture-independent* point of view conceivable, i.e., there is no external point of view which would enable a *direct* observation of *basic reality*. Thus, questions like "How is *basic reality* built up and what kind of processes take place in it?" makes no sense. Therefore, the question "Is *Mach´s principle* satisfied within *basic reality*?" also makes no sense.

In summary, *Mach´s principle* is satisfied within the frame of the *principle of level-analysis* without any restriction.

Space-time only arises with objects and processes

The picture and also its frame, space and time, are located in the head of the observer. We know from experience that space-time arises only in connection with objects and processes, i.e., an *empty* space-time is not observable.

That obviously means that space-time is not installed in the brain as a definite unit but is only "created" if there is actually something for representation, i.e., when our sense-organs take up objects and processes from reality outside. In this way we can explain the *test-body effect* discussed in Subsec. 1.1.6 (distances and time intervals can only be measured in connection with material objects and physical processes, respectively).

Space and time are exclusively used for the construction of pictures of reality, i.e., space and time become superfluous if there is nothing for representation; we have recognized in Subsec. 1.3 that superfluous things had no place within the process of evolution. In conclusion, also from the point of view of evolution, space and time should only appear if there is actually something for representation.

Furthermore, objects are ordered in space and time *relative to each other*, i.e., it makes no sense to order a *single* object and this explains that only distances are observable and not single space-time points.

All these problems, which are obviously insuperable within the frame of the usual conception (details are given in Subsec. 1.1), are solved within the *principle of level-analysis* in the most natural way; space and time are elements of the *cognition apparatus* of the observer and this apparatus has been developed in accordance with evolution.

Chapter 2

The Point of View of Philosophy of Science

2.1 The Asymptotic Convergentism

2.1.1 Conclusions from the Structure of Scientific Theories

In accordance with the *principle of level-analysis*, the concept of *basic reality* is of certain relevance. However, *basic reality* is principally not accessible to the observer, i.e., we cannot make any direct statement about it. But we can construct *pictures* of *basic reality* on various *levels*. Here we can distinguish between two possibilities:

1. The pictures are formed in an *unconscious* way by our sense organs and our cognition system, i.e., the pictures come into play *without* the conscious action of the observer.

2. The pictures are constructed by *thinking*, i.e., the pictures come into play *with the conscious* action of the observer.

The *levels of reality* are arranged hierarchically in accordance with the degree of generality. Furthermore, from the *principle of level-analysis*, it follows that there can be no similarity between the pictures and the actual compositions in *basic reality*. All these things have been discussed in detail above.

It is interesting to note that this "conception of the world" is supported by the *philisophy of science*; in this chapter we would like to discuss this point in more detail.

A fundamental question

A representative question, which is treated in the philosophy of science, can be formulated as follows [19]:

Are electrons, quarks or other physical entities really existing units? In other words, are they units of "reality in itself". The term "reality in itself" is understood to mean a single defined complex independent of us (i.e., independent of the intellect, the language, our scheme of notations and theories)?

In this sense the concept "reality in itself" is identical with the concept "basic reality", which we have introduced above. However, in this chapter we want to discuss certain facts in connection with the philosophy of science independent of the *principle of level-analysis* and, therefore, we do not want to use the concepts at first, which we have deduced in connection with the *principle of level-analysis*. At the end of this chapter comparisons will be made.

From the point of view of the *principle of level-analysis* the question (formulated above) "Are electrons, quarks or other physical entities really existing units?" has to be answered in the negative. However, not only on the basis of this principle but also from the point of view of philosophy of science, at any rate when we assume that *Kuhn's theses* are correct [28]; according to that we have to seriously doubt that we can make statements about "reality in itself", and these doubts can be well-founded in a convincing manner by analyses within the philosophy of science.

Only on the basis of a scientific theory can we define the entities which do exist and those which do not and how they behave. Then the question of whether the world actually is what science considers it to be implies studying the problem of whether scientific theories are *really true* or not.

In this connection it is interesting to note that our notions concerning reality are obviously not only *supplemented* in the course of time but *reformulated on the basis of new first principles.* This will be investigated in more detail in the following sections on the basis of philosophy of science.

On the generality of the principle of level-analysis

In conclusion, quite independent of the remarks and analyses of the preceding sections, arguments from the philosophy of science lead to features which are commensurable with those obtained from the *principle of level-analysis*. In other words, from the structure of scientific theories it can be concluded that the *principle of level-analysis* is realized in nature or at least a principle which is closely related to it. Since the investigations concerning the structure of scientific theories are very general, we may also conclude that the *principle of level-analysis* is a far-reaching principle in nature, i.e., *all* observed phenomena should be based on it.

As already remarked above, the arguments which we will use in this chapter are essentially based on the theses by *Thomas Kuhn* which he published in 1962 as a monograph entitled, "The Stucture of Scientific Revolutions" [28]. In this connection we will be interested in the question of whether science progresses by *successive refinement* of a basic conception or by *substitution* of a conception by another which is fundamentally different from the preceding one. In this section we will first investigate the view that science progresses by *successive refinement (asymptotic convergentism)*.

2.1.2 Tendencies in the Past and in the Present

Self-indulgence is dominating

How does science progress? Is there anything like a "final view" of the things? Which tendencies exist historically to categorize a status of knowledge acquired? For instance, *Lord Kelvin* (1824-1907) thought that the foundations of physics as laid down towards the end of the last century were complete and that only secondary questions were still left to be answered. *Berthelot* (1827-1907) in 1885 felt that the world no longer concealed any secrets. *Haeckel* (1834-1919) concluded from his studies (also made towards the end of the 19th century) that all legitimate questions in natural science had essentially been answered. Another example of this tendency is given by *Max Planck* (1859-1947) [19]:

"As I was beginning to study physics (in 1875) and sought advice regarding the conditions and prospects of my studies from my eminent teacher Philipp von Jolly, he depicted physics as a highly developed and virtually full-grown science, which - since the discovery of the principle of the conservation of energy had in a certain sense put the keystone in place - would soon assume its final stable form. Perhaps in this or that corner there would still be some minor detail to check out and coordinate, but the system as a whole stood relatively secure, and theoretical physics was markedly approaching that the degree of completeness which geometry, for example, had already achieved for hundreds of years. Fifty years ago (as of 1824) this was the view of a physicist who stood at the pinnacle of the times."

If we jump from the last century to the present, we find a recurrence of the opinions above. For instance, Richard Feynman wrote [19,29]:

"What of the future of this adventure? What will happen ultimately? We are going along guessing the laws; how many laws are we going to have to guess? I do not know. Some of my colleagues say that this fundamental aspect of our science will go on; but I think there will certainly not be perpetual novelty, say for a thousand years. This thing cannot keep on going so that we are always going to discover more and more new laws ...We are very lucky to live in an age in which we are still making discoveries. It is like the

discovery of America - you only discover it once. The age in which we live is the age in which we are discovering the fundamental laws of nature, and that day will never come again. It is very exciting, it is marvellous, but this excitement will have to go. Of course in the future there will be other interests,... There will be a degeneration of ideas, just like the degeneration that great explorers feel is occuring when tourists begin moving in on a territory. In this (present) age people are (perhaps for the last time?) experiencing a delight, the tremendous delight that you get when you guess how nature will work in a situation never seen before."

It is clearly evident from all the examples cited that eminent personalities (and obviously not only these) in different epochs thought that scientific findings "now" approached their final state of completion. Evidently, there have also been epochs in which the knowledge available was judged with less self-indulgence. *Nicholas Rescher* in his book entitled, "The Limits of Science" [19], from which most of the texts of the examples quoted above have been taken, deals with these antipositions. However, compared with the discussions in present day literature, the opinions of the authors quoted before - especially that expressed by Feynman - seem to dominate for the time being.

2.1.3 How Does Science Progress?

Generally speaking, all these concepts quoted above seem to have in common the following basic principle:

Science progresses by eliminating the number of unanswered questions.

Let us suppose that there exists a defined set R_S of problems which, in the course of time, are solved, i.e., the questions asked are answered successively. If we denote the number of *answered* questions at time t

by R and the number of *answered* questions at time t' ($t'>t$) by R', we obtain (see Fig. 23)

$$R < R' \ . \tag{25}$$

Accordingly, the number of questions capable of formulation decreases:

$$(R_s - R) > (R_s - R') \ . \tag{26}$$

Within the frame of such a principle *all* questions will be answered in the course of time, i.e., a maximum of scientific knowledge is achieved.

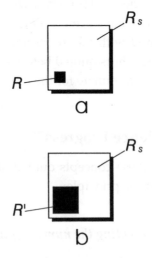

Figure 23. a: Knowledge at time t: There is a constant number of R_S problems, the number of answered questions is R. b: Knowledge at time $t' > t$: There are R_S problems, the number of answered questions is $R' > R$.

Principle of propagation of questions

In contrast to this, *Immanuel Kant* (1724-1804) advocates the "Principle of Propagation of Question ". It means that each answer to a scientific question gives rise to *new* questions. According to Kant this applies because each answer provides new material feeding new questions. But new questions generally change the supposition and this may lead to an *extension of the problem horizon*: Instead of looking at Fig. 23, we are induced to read Fig. 24. Now a variable (i.e., time dependent) set of problems is encountered. Then an increase in the number of unanswered questions can very well accompany an increase in answered questions. Since

$$R_S < R'_S \tag{27}$$

(see Fig. 24) the following relations

$$R < R'$$
$$(R_S - R) < (R'_S - R') \tag{28}$$

are possible.

However, the opinions of the scientists quoted above (Lord Kelvin, Feynman, etc.) can be well harmonized with Kant's "Principle of question propagation" provided that not only the number of questions but also their *relevance* is taken into consideration. As a matter of fact, if the relevance of a problem gradually decreases in the course of time, "later" science must be less important. In that case, it can be quite correctly asserted that the basic structure of physics has been worked out although the number of questions to be answered increases with time.

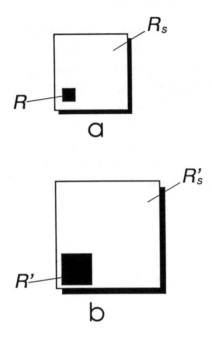

Figure 24. a: Knowledge at time t: There are R_s problems, the number of answered questions is R. b: Knowledge at time $t' > t$: There are R_s' problems, the number of answered questions is R'. The number of unanswered questions increases with time.

Such a view of the progress of science suggests the analogy with geographic research. In this context, Rescher states among others [19]:

"Scientific inquiry would thus be conceived of as analogous to terrestrial exploration, whose product - geography - yields results of continually smaller significance which fill in ever more minute gaps in our information. In such a view, later investigations yield findings of ever smaller importance, with each successive accreation making a relatively smaller contribution to what has already come to hand. The advance of science leads, step by diminished

step, toward a fixed and final view of things."

Accordingly, science progresses by successively approaching the truth: The "final answer" and the "final view" of things, respectively, are gradually approaching the final view (truth) by the way of an asymptotic approximation. According to *Peirce* (1839-1914) this truth is attained in the limit $t \mapsto \infty$:

$$R_\infty = \lim_{t \to \infty} R_t \; , \tag{29}$$

where R_t is the time dependent status of knowledge which approaches asymptotically the definite truth R. This detailed filling (*accumulation*) of given, fundamentally defined patterns resembles greatly the calculation of further decimal points in order to additionally refine a result already roughly estimated, such as in calculating the numerical value of π:

$$\pi_1 = 3.1$$
$$\pi_2 = 3.14$$
$$\pi_3 = 3.141$$
$$\vdots$$
$$\pi = \lim_{n \to \infty} \pi_n \; . \tag{30}$$

At least until one generation ago, the opinion was firmly established that science is *cumulative* and the advocates of the scientific method tried to understand scientific progress to have this cumulative nature (see also [19]). Within this concept the "real truth" is set equal to "our ultimate truth". Accordingly, the question asked at the beginning of whether physical entities (electrons, quarks, etc.) do really exist must be answered in the positive. But we will see below that the *asymptotic convergentism* cannot be upheld in the recognition theory and scientific history.

2.2 Substitution Instead of Successive Refinement

2.2.1 Critical Remarks to the Asymptotic Convergentism

Essentially, two serious objections can be advanced against the statements in Subsec. 2.1, i.e., against the *asymptotic convergentism*:

1.There is no Metric to Measure the "Distance" between
  Bodies of Knowledge

The asymptotic convergentism is from the start burdened by the great difficulty that it cannot give a metric system allowing us to define the interval concerned

$$R_t - R_{t'}$$

between the status of knowledge R_t and the status of knowledge $R_{t'}$. This means that we are not at all in a position to decide whether we have or have not approached the "real truth". How can criteria be formulated according to which such an approximation could be measured? There is quite simply no neutral point of view in theoretical terms, i.e., no neutral, external to science, elevated level which could form the basis for a direct comparison between theoretical configurations R_t and $R_{t'}$, etc. Scientific progress can solely be measured on the so-called *pragmatic* level. This point will be discussed in more detail below.

2. There is a fundamental change of mind

The assumption of a successive approximation cannot be maintained in view of the history of science because the analysis of theories succeeding each other in time shows that the later theory is generally not only *supplemented* and *refined*, respectively, but *reformulated on the basis of new first principles*. A basic *change in perspective* takes place. Normally the problem does not consist of just adding some further facts but of structuring a *new frame of thinking*.

Discussion

This situation has been described in a highly instructive manner by Kuhn who compared *Newton's theory* to the *theory of relativity* [28]:

"From the viewpoint of this essay these two theories are fundamentally incompatible in the sense illustrated by the relation of Copernican to Ptolemaic astronomy: Einstein's theory can be accepted only with the recognition that Newton's was wrong. Today, this remains a minority view. We must therefore examine the most prevalent objections to it.

The gist of these objections can be developed as follows. Relativistic dynamics cannot have shown Newtonian dynamics to be wrong, for Newtonian dynamics is still used with great success by most engineers and, in selected applications, by many physicists. Furthermore, the propriety of this use of the older theory can be proved from the very theory that has, in other applications, replaced it. Einstein's theory can be used to show that predictions from Newton's equations will be as good as our measuring instruments in all applications that satisfy a small number of restrictive conditions. For example, if Newton's theory is to provide a good approximative solution, the relative velocities of the bodies considered must be small compared with the velocity of light. Subject of this condition and a few others, Newtonian theory seems to be derivable from Einsteinian, of which it is therefore a special case.

But, the objection continues, no theory can possibly conflict with one of its special cases. If Einsteinian science seems to make Newtonian dynamics wrong, that is only because some Newtonians were so incautious as to claim that Newtonian theory yielded entirely precise results or that it was valid at very high relative velocities. Since they could not have had any evidence for such claims, they betrayed the standards of science when they made them. In so far as Newtonian theory was ever a truly scientific theory supported by valid evidence, it still is. Only extravagant claims for the theory - claims that were never properly parts of science - can have been shown by Einstein to be wrong. Purged of these merely human extravagances, Newtonian theory has never been challenged and cannot be."

Kuhn adds further comments on this situation as follows [28]:

"Can Newtonian dynamics really be derived from relativistic dynamics? What would such a derivation look like? Imagine a set of statements

$$E_1, E_2, ..., E_n,$$

which together embody the laws of relativity theory. These statements contain variables and parameters representing spatial position, time, rest mass, etc. From them, together with the apparatus of logic and mathematics, is deducible a whole set of further statements including some that can be checked by observation. To prove the adequacy of Newtonian dynamics as a special case, we must add to the E_i's additional statements, like

$$(v/c)^2 \ll 1,$$

restricting the range of the parameters and variables. This enlarged set of statements is then manipulated to yield a new set,

$$N_1, N_2, ..., N_m,$$

which is identical in form with Newton's laws of motion, the laws of gravity, and so on. Apparently Newtonian dynamics has been derived from Einsteinian, subject to a few limiting conditions.

Yet the derivation is spurious, at least to this point. Though the N_i's are a special case of the laws of relativistic mechanics, they are not Newton's

Laws. Or at least they are not unless those laws are reinterpreted in a way that would have been impossible until after Einstein's work. The variables and parameters that in the Einsteinian E_i's represented spatial position, time, mass, etc., still occur in the N_i's; and they still represent Einsteinian space, time, and mass. But the physical referents of these Einsteinian concepts are by no means identical with those of the Newtonian concepts that bear the same name. (Newtonian mass is conserved; Einsteinian is convertible with energy. Only at low relative velocities may the two be measured in the same way, and even then they must not be conceived to be the same.) Unless we change the definitions of the variables in the N_i's, the statements we have derived are not Newtonian. If we do not change them, we cannot properly be said to have derived Newton's Laws, at least not in any sense of "derive" now generally recognized."

This implies that Einstein's theory, as compared to Newton's theory, provides a basically novel frame of thinking. A change in perspective has taken place, involving the *notational decoupling* and *displacement*, respectively, of the network of notions. Newton's theory has proved to be too coarse in its fundamental concept. It is obvious from the quotation above that in Kuhn's opinion Newton's theory has proved to be *erroneous*. We think that this formulation is too harsh. This point will be dicussed in more detail in Subsec. 2.3.

2.2.2 Can We Say something about the "Absolute Truth"?

Estimates of the truth

Thus, generally speaking, we have to conclude that a "later" theory became necessary because an "earlier" theory had been limited in its scope. This led to a basically novel concept of the way of seeing nature. Thus, normally, not only improvements and refinements, respectively, are made, but the "earlier" theory is downright replaced by the "later" one. The history of science provides a wealth of examples supporting this statement (see also [19, 28]). Thus, we have to assume consistently that the "later" theory will also have to be abandoned at some point in time. So each frame of thinking, independent of the era in which it has been conceived, can never constitute a frame for the "real (absolute) truth". In this context we cannot even provide evidence that we have come closer to the "absolute truth" because no metric system can be defined to measure intervals in recognition. A framework of thinking and a theory, respectively, do not reflect the "real truth" but - as formulated by Rescher [19] - an "estimate of the truth" which is to be understood as meaning a tentatively postulated provisional truth. This point will not be further treated here, but dealt with in more

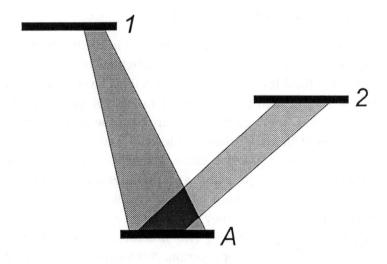

Figure 25

detail in Subsec. 2.3. Here, we will be elucidate what we mean when we speak of "scientific progress".

Incommensurable structures of thinking

When changing from one to another frame of thinking, a conceptional decoupling generally takes place so that successive structures of thinking may become *incommensurable*. The advocates of incommensurable theories basically are not in a position to understand each other, because it is not reasonable to make comparisons between incommensurable structures of thinking. Using the following analogy, Rescher reduces the problem to the point [19]: *"One can improve upon one´s car by getting a better car, but one cannot improve it by getting a computer or a dishwashing machine."*

2.2.3 The "Pragmatic" Level

Scientific progress can be defined only if it is possible to project certain tendencies of two incommensurable structures of thinking to one "appropriate" third level (see Fig. 25). This third level which then is a sort of reference system (this is level A in Fig. 25) generally will have a "coarser" and "more global" structure than the two incommensurable structures of thinking (these are levels 1 and 2 in Fig. 25). They are coarser and more global because, generally, such a projection is not detailed (i.e., point by point) and provides only an integral picture (e.g., by averaging). These integral variables of levels 1 and 2 have things in common, if they cover a joint zone on level A. Level A is "appropriate" for comparing two structures of thinking, if it offers to them a finite surface for projection. A structure of thinking is superior to another if it describes in more detail and accuracy the body of facts on level A. In this way it is possible to compare (albeit to a limited extent only) two competing theories. In Fig. 25, the structure of thinking underlying level 1 is superior to that of level 2 because the surface (equivalent to the status of recognition which, starting from level 1, explores level A) of level 1 projected onto level A is larger than the surface onto level 2 projected onto level A.

Related to the situation of man, level A might constitute the level of everyday life, of technological applications, and of experimental explorations (in Chapter 1 we have called this level "level of measuring instruments"). Thus, technological progress and the elucidation by experiments become the touchstone (albeit to a limited extent) of deviating theoretical positions. Applying this yardstick, a "later" theory must preserve and improve the practical successes of its predecessors, it is then superior to the "earlier" theories. But it must always be kept in mind that the judgement of a given structure of thinking from the "pragmatic" level provides but a restricted perspective.

Scientific progress defined in this way will depend in many domains essentially on the technological progress because natural science often needs a more sophisticated technology in order to be able to perform its increasingly complicated interactions with nature. According to what has been said before, from the cognitive point of view, natural science repeatedly starts from its origins; however, given the limits imposed by technology, this process, for merely *practical* reasons, will proceed at an ever slower rate.

2.3 Pictures of "Reality in itself"

Summary

Let us summarize once more the major items of Subsecs. 2.1 and 2.2:
According to the *asymptotic convergentism*, science progresses by successive approximation to the "final answer" which is set equal to the "real (absolute) truth". But it has appeared that this notion cannot be maintained within the framework of the theory of recognition and the history of science [19, 28]. The reason is that there is no refinement or improvement but a *substitution* of models of thinking. Theoretical scientists (among others Feyerabend, Kuhn, Lakatos, Quine and Rescher) of the most diverse disciplines agree that the traditional theory of accumulation does not work. *Karl Popper* states in this context [30]:

"It is not the accumulation of observations which I have in mind when I speak of the growth of scientific knowledge, but the repeated overthrow of scientific theories, and their replacement by better or more satisfactory ones."

It has to be considered as an *empirical* fact that a great pace in natural science is always connected with a basic change in perspective, and the "later" frame of thinking has but little or nothing in common with the "earlier" one. In other words, while such transitions take place, there are corrections down to the first principles. Various schemes of thinking are generally *incommensurable*. That means that it is not meaningful to make direct comparisons.

Scientific progress can be defined to a limited extent only on the "pragmatic" level; a *continuous* progress in technology is normally preceded by a *discontinous* change in the structures of thinking because in terms of cognition there is a "permanent restart" (see also [19, 28]).

2.3.1 Consequences

Things come and go

Newton´s theory, as rightly stated by Kuhn, cannot be derived from the theory of relativity. Kuhn is also right when he says that the advocates of incommensurable theories (such as Newton´s theory and the theory of relativity) are hardly in a position to understand each other. However, in that case the advocate of the relativity theory strictly speaking cannot assert that Newton´s theory is *erroneous*. He can only assert that the theory of relativity is superior to Newton´s theory at the "pragmatic" level. But under a certain aspect each of the two theories is an "estimate of the truth". The theory of relativity covers a larger range of the "pragmatic" level and hence providing a better estimation. This means that each frame of thinking constitutes a better or worse "estimation of the truth". Thus, also the precopernian conception of the world, seen under a certain point of view, has its justification. Let us quote Kuhn in this context [28]:

"Communication across the revolutionary divide is inevitably partial. Consider, for another example, the men who called Copernicus mad because he proclaimed that the earth moved. They were not either just wrong or quite wrong. Part of what they meant by "earth" was a fixed position. Their earth, at least, could not be moved. Correspondingly, Copernicus´ innovation was not simply to move the earth. Rather, it was a whole new way of regarding the problems of physics and astronomy, one that necessarily changed the meaning of both "earth" and "motion". Without those changes, the concept of a moving earth was mad."

Each "estimation of the truth" is generally notationally decoupled from the other or a fundamental displacement has occured in the under-standing of notations. Such a process includes that, generally, certain terms (electrons, quarks, etc.) may even disappear while others appear. Not integrated into this sequence of "estimations of the truth" are *false* theories. The history of science shows that there is a wealth of such concepts. For instance, the theory of a light conducting aether has proved to be wrong; it was eliminated.

Because of the "permanent restart" from the cognitive point of view we cannot assume that we have just now arrived at a cognitive end point. We must rather suppose that in the future a completely new frame of thinking might be necessary or appear which makes us see things and reality, respectively, in quite a different light. On the other hand, we have to take for granted that the fundamental reality ("reality in itself") does not undergo changes during these processes.

The fundamental reality from the point of view of the theory of science

From our discussion it follows that two points should be considered as established:

1. There is a hierarchy ("depth") of generally incommensurable "estimations of the truth".

2. The world itself does not change when the frame of thinking is changed.

Both points are only compatible with each other if the world, i.e., the fundamental reality, is described by *pictures*. These pictures constitute the "estimations of the truth". Each picture has its justification under certain experimental suppostions and a conceptional frame. Those pictures give a *deeper* representation of the fundamental reality which gives a more comprehensive and more accurate reproduction of the "pragmatic" level. But on account of their incommensurability, these pictures cannot be precise reproductions of the fundamental reality, i.e., the theoretical terms (electrons, quarks, etc.) of natural science cannot refer to actually existing entities because we cannot know, from the point of view of the theory of science, the "real (absolute) truth" but only "estimations of the truth". (This is of course also the case when we analyse the world on the basis of the *principle of level-analysis*, which we have introduced in Chapter 1.)

In other words, the world cannot be as envisaged by science; thus, the picture itself does not resemble the fundamental reality because successive schemes of thinking generally correspond to basically differing pictures, although the world has not undergone changes.

Within the *asymptotic convergentism* it can rather be assumed that the world is actually as science envisages it to be. Here, fundamentally new pictures of the world are not generated again and again, but *one* frame is filled successively. In that case the concept is justified that theoretical terms like electrons, quarks, etc. are actually existing entities in the world. However, as already outlined in detail above, the asymptotic convergentism can no longer be upheld. The *empirical* finding (resulting from the analysis of facts provided by the history of science) that science progresses by a sequence of incommensurable schemes of thinking (pictures) must be given a rank equal to that of a relevant experiment in a laboratory.

2.3.2 Connections to the Principle of Level-Analysis

From the structure of successive schemes of thinking (theories) we can draw conclusions about "reality in itself". One of the essential results was that we cannot know the "absolute truth" because we are not able to define a *metric* to measure the "distance" between bodies of knowledge. This shortcoming indicates what kind of statements can be made by man concerning "reality in itself": From the point of view of the philosophy of science, "reality in itself" cannot be *directly* registered, but we can form *pictures* of it and these pictures are identical with Rescher´s "estimations of the truth". These pictures or "estimations of the truth" does not reflect the "reality in itself" in the form of a precise reproduction, i.e., the theoretical terms (electrons, quarks, etc.) in physics are by no means really existing entities; such terms can disappear while others can appear, i.e., things come and go.

These features, deduced from the philosophy of science, are identical with those which we have found in Chapter 1 on the basis of the *principle of reality-analysis*: "Reality in itself" (fundamental reality) is identical with *basic reality*, the *pictures of (basic) reality* can be equated with Rescher´s "estimations of the truth" which we have also called pictures.

The "estimations of the truth" can be hierarchically ordered on the basis of a reference level. In Subsec. 2.2.3, we have used the "pragmatic" level as a reference level. Then - so we have argued - a certain structure of thinking (theory) is superior to another if it describes the facts at the "pragmatic" level more extensively and more precisely. A theory which is superior (i.e., more general) to another can be formally arranged above the other and in such a way we obtain a hierarchy which is identical to that which we have formulated within the *principle of level-analysis*: The levels are ordered in accordance with the degree of generality (see also Fig. 10). The "pragmatic" level must be identified with the *level of everyday life* as well as with the *level of measuring instruments* which have been introduced in Chapter 1. However, within the frame of the *principle of level-analysis* the "pragmatic" level (level of everyday life, level of measuring instruments) does not play that outstanding role as within the frame of conventional physics because, within the frame of the *principle of level-analysis*, all levels are similar in character (see in particular Subsec. 1.4.3), i.e., in principle we could also choose other reference levels for assessment, and we would come to other criteria of assessment. In other words, structures of thinking (theories) must not exclusively be tested on the basis of experimental data, but also by means of other theories (test of a theory by a theory). Such a procedure may possibly be dominant in future, since the development of experimental devices may sooner or later reach inherent limits.

In conclusion, considerations on the basis of the philosophy of science support impressively the *principle of level-analysis*; we may state that the *principle of level-analysis* is generalized in a certain sense by the results of the philosophy of science since these results are very general.

2.3.3 Do Unconscious Changes in Perspective Exist?

For the assessment of an experimental situation a framework of thinking is needed. On the basis of such a framework, a picture of reality can be modeled which should be compatible with the experimental data. *"Looking at a bubble-chamber photograph, the student sees confused and broken lines, the physicist a record of familiar subnuclear events"* [28], i.e., no assessment of a certain visual impression (bubble-chamber photograph) was possible without a framework of thinking.

As is well known, that in the case of a changed experimental situation, it may occur that the present frame of thinking does no longer lead to pictures of reality which are compatible with the experimental data, produced by the changed experimental situation. In such cases we have to change the frame of thinking: The new frame of thinking is generally not only *supplemented* and *refined*, respectively, but *reformulated on the basis of new first principles*.

The creation of a frame of thinking is based on a process of thinking; it defines a certain level of reality and can be applied to other (lower) levels, e.g., the level of measuring instruments (Subsec. 1.5.1). As we have already discussed in detail in Subsec. 1.4.5, thinking is a *conscious* process, i.e., the process of thinking takes place *with* conscious actions of the observer. In particular, we have outlined in Subsec. 1.4.5 that thinking is characterized by *conscious objectivation*, in conrrast to *unconscious objectivation* which takes place without conscious actions of the observer. The transition from one frame of thinking to another means to carry out a "conscious change in perspective".

According to the *principle of level-analysis*, conscious and unconscious objectivation are processes which are similar in character; in both cases pictures of one and the same reality (basic reality) are produced, i.e., by conscious and unconscious objectivation, respectively, pictures come into being that are merely arranged on different *levels of reality*. This fact has been well-founded in Subsec. 1.4.3-1.4.5; we have stated there that the similarity between conscious and unconscious objectivation is due to mechanisms (*constancy mechanisms*, see Subsec. 1.4.5) which are similar in character.

If the similarity between conscious and unconscious objectivation is actually a general phenomenon, we also have to assume that a certain kind of "unconscious change in perspective" should happen in nature, i.e., a "change in perspective" should be possible *without* the conscious action of the observer. Such processes actually take place and the following remarks by Kuhn support this thesis [28]:

"An experimental subject who puts on goggles fitted with inverting lenses intially sees the entire world upside down. At the start his perceptual apparatus functions as it had been trained to function in the absence of goggles, and the result is extreme disorientation, an acute personal crisis. But after the subject has begun to learn to deal with his new world, his entire visual

field flips over, usually after an inverting period in which vision is simply confused. Thereafter, objects are again seen as they had been before the goggles were put on. The assimilation of a previously anomalous visual field has reacted upon and changed the field itself. Literally as well as metaphorically, the man accustomed to inverting lenses has undergone a revolutionary transformation of vision."

In conclusion, a revolutionary transformation took place without the conscious action of the observer. In other words, an "unconscious change in perspective" came into play. In this connection, let us quote a further interesting statement by Kuhn [28]:

"Surveying the rich experimental literature from which these examples are drawn makes one suspect that something like a paradigm is a prerequisite to perception itself. What a man sees depends both upon what he looks at and also upon what his previous visual-conceptual experience has taught him to see. In the absence of such training there can only be, in William James' phrase, < a bloomin´ buzzin´ confusion >."

According to the *principle of level-analysis,* we can understand these facts in a relatively simple way: The external stimuli do not yet contain a picture of the outside world, since the picture will only be produced in the head of the observer. The production of a picture is necessary because there is no similarity between the picture and the outside world, so as the cinema ticket has no similarity with the cinema itself (see also Subsec. 1.3.3). The picture in the experiment with inverting lenses is produced *unconsciously,* i.e., here the cognition apparatus models *unconsciously* - as in all the other cases within assumptionless observations. But we have learned from the experiment with inverting lenses, that the cognition apparatus models after a certain time in another way than in the case without inverting lenses (" ... *the inverting lenses show that two men with different retinal impressions can see the same thing."* [28]). This "change in perspective" takes place without the conscious action of the observer, i.e., we can state that there are "unconscious changes in perspective" in nature which can actually be undertood only on the basis of the *principle of level-analysis*.

Epilogue

The big success of Newton´s mechanics established a world view, which, at the height of his success, approximately at the end of the nineteenth century, became a model for all the other sciences. So, in many instances, even living human bodies have been considered as a mechanical machine.

With the discovery of quantum phenomena this view has been substantially modified, but not at all completely. In the formulation of the quantum mechanical apparatus, some terms and conceptions, which were the typical features of classical mechanics, slipped into the theory, i.e., in the formulation of quantum theory, it was evidently not achieved to completely leave Newton´s mechanics; in our opinion, this is reflected by the fact that there does not exist a satisfactory interpretation of quantum theory up to the present day, although the essential features of the mathematical formalism of (non-relativistic) quantum theory were already constructed by *Werner Heisenberg* (1901-1976) and *Erwin Schrödinger* (1887-1961) in 1925-1926. There exist equivalent pictures of reality, which are, on the other hand, completely different from each other, and they are strange and bizarre. These worlds and the mechanistic world view, respectively, have among other things the following common feature:

Reality is embedded in space and time.

But this conception becomes questionable when we consider the facts which we have discussed in this monograph, and these facts indicate that the relationship between reality and space-time have to be arranged in another way:

Reality should be projected onto space-time.

This conception is supported by new investigations. What are these new investigations? We are concerned with experimental results which show that there must exist *synchronous events* in physical reality. These are events which take place simultaneously at different positions in space and which are connected to each other by a meaningful features as well. Not only in physics but also in psychology, biology, behavioural research, etc., synchronous events have obviously been observed.

In the opinion of many scientists, new ways have to be gone in order to understand the effect of synchronicity, ways which leave the path of conventional quantum theory. For example, *David Bohm* regards the world as a hologram embedded in high-dimensional space. In [5], we have taken the view that the relationship between reality and space-time should be different from that used in conventional theories; in particular, we have used the conception which has been well established in this monograph, i.e., that reality is not *embedded in* space and time but that reality is *projected onto* space-time.

We have shown in [5] that such a rearrangement is very helpful in the solution of some relevant problems, problems which have caused considerable problems in the usual quantum theory. These problems particularly appear in connection with the terms "time" and "particle".

In a forthcoming work [31] the foundations of quantum theory and classical mechanics will be discussed in connection with the new principles. This will be done in more detail than in the book "Quantum Theory and Pictures of Reality" (Chapter 5) [5]. New terms such as "basic reality", "level of reality", etc., which we have introduced in this monograph, will be used in argumentations and operations.

Furthermore, we will emphasize in [31] that in the fundamental tendencies of our results, typical features are also reflected which are relevant in connection with other areas in science as psychology, biology and the philosophy of science.

References

1   Ivar Ekeland, *Mathematics and the Unexpected*, University of Chicago Press, Chicago and London, 1988.

2   C. G. Jung, *Synchronismus, Akausalität und Okkultismus*, Deutscher Taschenbuch Verlag GmbH & Co. KG, München, 1990.

3   R. Havemann, *Dialektik ohne Dogma?*, Rowohlt Taschenbuch Verlag GmbH, Reinbek bei Hamburg, 1964.

4   G. Falk, *PHYSIK, Zahl und Realität*, Birkhäuser Verlag, Basel, Boston, Berlin, 1990.

5   W. Schommers (Ed.), *Quantum Theory and Pictures of Reality, Foundations, Interpretations, and New Aspects*. With Contributions by B.d'Espagnat, P. Eberhard, W. Schommers, F. Selleri, Springer-Verlag, Berlin, Heidelberg, 1989.
    Reprinted by World Publishing Corporation, Beijing, Volksrepublik China, 1990.

6   A. Einstein, *Grundzüge der Relativitätstheorie*, Verlag Vieweg und Sohn, Braunschweig, 1973.

7   M. Born, *Die Relativitätstheorie Einsteins*, Springer-Verlag, Berlin, Heidelberg, 1964.

8   B. Hoffmann (with the collaboration of H. Dukas), *Albert Einstein - Creator and Rebel*, The Viking Press, New York, 1972.

9   K. Gödel, *An Example of a New Type of Cosmological Solutions of*

Einstein's Field Equations of Gravitation, Rev. Mod. Phys. **21**, 447, 1947.

10 O. Heckmann, *Sterne, Kosmos, Weltmodelle*, Deutscher Taschenbuch Verlag Verlag GmbH & Co. KG, München, 1980.

11 B. Kanitscheider, *Kosmologie*, Philipp Reclam jun., Stuttgart, 1984.

12 H. Dehnen, in *Philosophie und Physik der Raum-Zeit*, J. Audretsch und K. Mainzer (Hrgs.), Wissenschaftsverlag, Mann - heim , Wien, Zürich, 1988.

13 M. Berry, *Principles of Cosmology and Gravitation*, Adam Hilger, Bristol and Philadelphia, 1989.

14 R. und H. Sexl, *Weiße Zwerge - Schwarze Löcher*, Rowohlt Taschenbuch Verlag, Reinbek bei Hamburg, 1975.

15 G. Falk und W. Ruppel, *Mechanik, Relativität, Gravitation*, Springer-Verlag, Berlin, Heidelberg, 1983.

16 E. Dammann, *Erkenntnisse jenseits von Raum und Zeit*, Droemersche Verlagsanstalt Th. Knaur Nachf., München, 1990.

17 H. v. Ditfurth, *Der Geist fiel nicht vom Himmel*, Deutscher Taschenbuch Verlag, München, 1980.

18 R. Brückner, *Das schielende Kind*, Schwabe & Co. Verlag, Basel/Stuttgart, 1976.

19 Nicholas Rescher, *The Limits of Science*, University of California Press, Berkelay, Los Angeles, London, 1984.

20 Michael Talbot, *Mysticism and the New Physics*, Routledge & Kegan Paul, London and Henley, 1981.

155

21 W. Schneider, *Hypothese, Experiment, Theorie*, Sammlung Göschen, Walter de Gruyter, Berlin, New York, 1978.

22 Karl R. Popper/Franz Kreuzer, *Offene Gesellschaft - Offenes Universum*, Ein Gespräch über das Lebenswerk des Philosophen, Piper, München, 1986.

23 Konrad Lorenz, *Die Rückseite des Spiegels*, Piper, München, 1973.

24 A.S. Eddington, *The Nature of the Physical World*, University of Chicago Press, 1963.

25 L.D. Landau, E.M. Lifschitz, *Lehrbuch der Theoretischen Physik III*, Quantenmechanik, Akademie-Verlag, Berlin, 1966.

26 F. Selleri, *Die Debatte um die Quantentheorie*, Friedr. Viehweg & Sohn, Braunschweig/Wiesbaden, 1983.

27 Albert Haug, *Theoretische Festkörperphysik*, Band 1, Franz Deuticke, Wien, 1964.

28 Thomas S. Kuhn, *The Structure of Scientific Revolutions*, University of Chicago Press, Ltd, London, 1962.

29 Richard P. Feynman, *The Character of Physical Law*, M.I.T. Press, Massachusetts Institute of Technology, Cambridge, Massachusetts, and London, England, 1965.

30 Karl Popper, *Truth, Rationality and the Growth of Scientific Knowledge*, in: "Karl Popper, Conjectures and Refutations", New York, 1962.

31. W. Schommers, *Symbols, Pictures and Quantum Reality*. On the Theoretical Foundations of the Physical Universe (in preparation).

Index

DATE DUE

FEB 1 8 1997			
APR 2 0 1997			
JAN 2 9			
MAR 0 8 1998			
DEC 0 4 2001			
GAYLORD			PRINTED IN U.S.A.